THE AMPHIBIANS AND REPTILES
OF MICHOACÁN, MÉXICO

THE AMPHIBIANS AND REPTILES OF MICHOACÁN, MÉXICO

WILLIAM EDWARD DUELLMAN

ISBN: 978-81-947476-5-9

Published: -

LECTOR HOUSE LLP
E-MAIL: lectorpublishing@gmail.com

THE AMPHIBIANS AND REPTILES OF MICHOACÁN, MÉXICO

BY

WILLIAM E. DUELLMAN

CONTENTS

INTRODUCTION

For almost 30 years North American herpetologists have been making extensive collections of reptiles and amphibians in México. Some parts of the country, because of their accessibility, soon became relatively well known; other regions lying off the beaten path were bypassed or inadequately sampled. Principally in the last decade herpetologists have been entering regions from which no collections previously were available in an attempt to fill gaps in known distributions and to discover unknown species of animals. In 1950 Dr. Donald D. Brand led an exploration party from the University of Texas to the poorly explored and faunistically unknown region of southwestern Michoacán. James A. Peters accompanied Brand and collected amphibians and reptiles. In 1951 I welcomed the opportunity to accompany Brand on a second expedition to southwestern Michoacán. Such was the beginning of my interest in the herpetofauna of the region. I have been fortunate to return to Michoacán on four successive trips, all of which had as their purpose the accumulation of data on the herpetofauna that would result in a survey of the component species and an analysis of their distribution.

My original intention was to amplify Peters' (1954) study based on the collections made by him in 1950 and by me in 1951 in the Sierra de Coalcomán. But it soon became evident that in order to understand the relationships of the herpetofauna of the Sierra de Coalcomán, the species inhabiting the Tepalcatepec Valley and adjacent mountain ranges would have to be studied. In the course of making that study I examined all specimens from Michoacán already in museums.

There have been few detailed herpetofaunal studies in México. The first such study of any consequence was that by Bogert and Oliver (1945) on the herpetofauna of Sonora. In that paper the authors analyzed the fauna from a geographic view and showed the transition from tropical species in the southern part of the state to members of the Sonoran Desert assemblage to the north. Martin (1958) made a detailed study of the herpetofauna of the Gómez Farías region in southern Tamaulipas; he emphasized the ecological distribution of amphibians and reptiles in that region with special reference to cloud forests. Duellman (1958c) presented a preliminary geographic analysis of the herpetofauna of Colima with special reference to the continuity of the species inhabiting the lowlands. Zweifel (1960) discussed in detail the herpetofauna of the Tres Marías Islands and commented on the derivation of the fauna. Duellman (1960d) provided a detailed account of the geographic distribution of the amphibians known to occur in the lowlands of the Isthmus of Tehuantepec and attempted to account for the present patterns of distribution.

The present report is the first of two parts dealing with the herpetofauna of Michoacán. The purpose of this part is to present a full account of the species of amphibians and reptiles known to inhabit the state of Michoacán; the accounts of the species are accompanied by a brief description of the natural landscape and of the various assemblages of species comprising the major faunistic groups within the region. A gazetteer of collecting localities is appended. The second part of the study, now in preparation, deals with the ecological and historical geography of the herpetofauna. Since the present part will be of interest primarily to systematic herpetologists, I have decided to separate it from the more general material of interest to biogeographers.

One of the major problems that faces the worker undertaking a faunal study is the presence of species or genera of unsettled systematic status. My work in Michoacán has been no exception; fifteen separate studies were undertaken in an attempt to solve systematic problems in certain groups. Some systematic problems still remain but are of little consequence insofar as the entire faunal picture is concerned, or are so involved as to be impractical to undertake at this time. In accounts of species, such problems are mentioned in the hope that they will interest some worker who will be inclined to investigate them.

ACKNOWLEDGMENTS

While engaged in the study of the herpetofauna of Michoacán I have built up a debt of gratitude to many individuals, without whose aid my ambition to complete my study never would have been realized. I am especially grateful to those individuals who accompanied me in the field; Lee D. Beatty, Richard E. Etheridge, Carter R. Gilbert, Fred G. Thompson, Jerome Tulecke, and John Wellman offered stimulating companionship and valuable assistance. On many occasions they suffered hardships on behalf of my interests.

Studies of my own specimens have been augmented by material from other institutions. For permitting me to examine specimens in their care I am indebted to W. Frank Blair, Charles M. Bogert, Doris M. Cochran, William B. Davis, James R. Dixon, the late Emmett R. Dunn, Josef Eiselt, Alice G. C. Grandison, Norman Hartweg, Robert F. Inger, Arthur Loveridge, the late Karl P. Schmidt, Hobart M. Smith, Robert C. Stebbins, Margaret Storey, Edward H. Taylor, and Richard G. Zweifel.

Several people have aided me in the study of specimens and in the analysis of data; I am grateful to Donald D. Brand, who first introduced me to Michoacán; since that time I have benefited much from his knowledge of the area. James A. Peters provided me with essential information concerning his field work in southern Michoacán in 1950. James R. Dixon and Floyd L. Downs have permitted me to use freely the material and data that they accumulated in their recent field work in Michoacán. Norman E. Hartweg allowed me to use the specimens and data that he gathered in his survey of the herpetofauna in the region of Volcán Parícutin. L. C. Stuart, Charles F. Walker, and Richard G. Zweifel have helped in unraveling some of the systematic and distributional problems.

I am especially grateful to my wife, Ann, who for six months helped me track down elusive species and explore new areas. Furthermore, she has stimulated me to carry this study to completion.

Many people in Michoacán favored the field parties with quarters, transportation, and valuable information, which greatly facilitated the field work. In this respect I am especially indebted to Ingeniero Ruben Erbina of Ingenieros Civiles Asociados, who not only let us use his home as our headquarters, but through a letter of introduction gave us the "key" to southern Michoacán. Ingeniero Pedro Tonda aided us in Arteaga and San Salvador. Ingeniero Anastacio Peréz Alfaro of the Comisión Tepalcatepec in Uruapan provided the latest maps of southern Michoacán and much essential information pertaining to travel conditions in the area. Señor Nefty Mendoza gave us a home in Dos Aguas; this kindness allowed us to work in this interesting region during the height of the rainy season. Mr. and Mrs. Bob Thomas let us make use of their facilities at Hacienda Zirimícuaro. The naval officers at the Estación Marina at Playa Azul made pleasant what might have been a dreadful stay in that small coastal village. To the managers and pilots of Lineas Aereas Picho in Uruapan I owe special thanks for going out of their way on more than one occasion to transport a stranded snake-hunter. Throughout the months of field work beginning in 1955 I constantly have been aided by the authorities and workers of the Comisión Tepalcatepec, a subdivision of the Secretaria de Caminos y Obras Publicas, and of the private corporation, Ingenieros Civiles Asociados. Much of the field work in Michoacán was made possible only through the co-operation of the natives who supplied mules, acted as guides, and aided in the collection of specimens. I have learned a great deal from these people. They will never see this report. Their work as guides, muleteers, and collectors greatly assisted me with the mountains of equipment that had to be piled on the backs of scrawny mules for transportation to places where the natives seldom trod. Their efforts in behalf of Don Guillermo never will be forgotten; I extend an especially hearty *muchas gracias* to Benjamin, Ignacio, Jesús, Lorenzo, Mariano, and Remigio.

Much of the work on this report was done while I was associated with the Museum of Zoology at the University of Michigan. I thank Norman E. Hartweg and T. H. Hubbell for making available to me the facilities of the museum and for their numerous courtesies that aided me so much.

My field work in Michoacán was supported by the Museum of Zoology at the University of Michigan (1951), by the Horace H. Rackham School of Graduate Studies of the University of Michigan (1955), by the Penrose Fund of the American Philosophical Society (1956), by the Bache Fund of the National Academy of Sciences (1958), and by the University of Kansas Endowment Association (1960).

Permits for collecting specimens in México were provided by the Dirección General de Caza through the courtesy of Ing. Juan Lozano Franco and Luis Macías Arellano.

HISTORICAL ACCOUNT

Unlike many parts of southern México and northern Central America, Michoacán received no attention from the collecting expeditions of the European museums in the last century. The earliest known herpetological specimens from Michoacán were obtained by Louis John Xantus, who was appointed U. S. Consul to Colima in 1859. In April, 1863, Xantus collected at Volcán Jorullo in Michoacán; in April and May of the same year he collected along the coast of Michoacán between the Río Cachán and the Río Nexpa. His small collection of 19 extant specimens is in the United States National Museum. Alfredo Dugès, a resident of Guanajuato, México, made early contributions to the knowledge of the herpetofauna of Michoacán. In 1885 he described *Sonora michoacanensis*, and in 1891 he described *Eumeces altamirani*; from what is known of the distribution of these species, he probably had collected in the Tepalcatepec Valley. During their biological survey of México, Edward W. Nelson and Edward A. Goldman spent a limited amount of time in Michoacán in 1892 and again in 1903 and 1904. Most of their collecting was done on the plateau in the north-central part of the state; their collections are in the United States National Museum. While collecting fishes in southern México, Seth E. Meek obtained some amphibians and reptiles from Lago de Pátzcuaro in 1904; these are in the collections of the Chicago Natural History Museum. In 1908 Hans Gadow ventured into the then unexplored "tierra caliente" of the Balsas Valley and collected at Volcán Jorullo and other localities in the valley. Later in the same year he collected at Guayabo, San Salvador, and Arteaga in the Sierra de Coalcomán and at Buena Vista and Cofradía in the Tepalcatepec Valley. His collections were deposited in the British Museum (Natural History) and the Naturhistorisches Museum Wien.

The first thirty years of the present century saw little more field work in Michoacán. In the 1930's Edward H. Taylor and Hobart M. Smith collected throughout much of México. At various times they worked in Michoacán, principally along the road from México City to Guadalajara. In 1935 Hobart M. Smith spent a week at Hacienda El Sabino south of Uruapan; he revisited the locality again in 1936 and made a large and important collection of amphibians and reptiles from the upper limits of the arid tropical scrub forest in the Tepalcatepec Valley. Specimens collected by Smith and Taylor were incorporated into the Edward H. Taylor-Hobart M. Smith collection, which subsequently was deposited in part in the Museum of Natural History at the University of Illinois and in part in the Chicago Natural History Museum. In 1939 Hobart M. Smith collected at Pátzcuaro and between Uruapan and Apatzingán; these collections, made while he was a Walter Rathbone Bacon Scholar of the Smithsonian Institution, are deposited in the United States National Museum. In 1940 and 1941 Frederick A. Shannon, who was a member of the Hoogstraal Expeditions under the auspices of the Chicago Natural History Museum, collected on Cerro de Tancítaro and at Apatzingán; an account of the specimens collected there was published by Schmidt and Shannon (1947).

The eruption of Volcán Parícutin in February, 1943, attracted the attention of many biologists, a group of which from the Museum of Zoology at the University of Michigan collected in the Cordillera Volcánica in 1945 and 1947. The amphibians and reptiles were collected and studied by Norman E. Hartweg. In 1950 James A. Peters accompanied Donald D. Brand on a preliminary exploration of the western part of the Sierra de Coalcomán and adjacent Pacific coast of Michoacán; in the same year Peters collected also on the Mexican Plateau and at Volcán Jorullo. His specimens are in the Museum of Zoology at the University of Michigan. Since 1950 many biologists have collected in Michoacán in the course of work on certain groups of animals or in general surveys. In this way Raymond Alcorn, Robert W. Dickerman, James R. Dixon, Floyd L. Downs, Emmet T. Hooper, and Robert R. Miller have contributed to our knowledge of the herpetofauna.

As stated previously, my own field work in Michoacán began in 1951, when I accompanied Donald D. Brand on an exploring expedition to the southern part of the state. In that year a short time was spent on the Mexican Plateau, principally in the area around Lago de Cuitzeo, and at Volcán Jorullo. In July and August we made our headquarters at Coalcomán. From that town the field party travelled southward to Maruata on the Pacific coast and thence back over the mountains to Coalcomán. Later in that summer we travelled by mule from Coalcomán southeastward to the mouth of the Río Nexpa. In 1955, accompanied by Lee D. Beatty, Carter R. Gilbert, and Fred G. Thompson, I collected in the Tepalcatepec Valley and at Coalcomán. We made a mule trip from Coalcomán to Cerro de Barolosa, where we made the first collections from the pine-fir forests in the Sierra de Coalcomán. Later in the same summer Carter R. Gilbert and I spent a week at Playa Azul on the Pacific coast. In March, April, and May, 1956, my wife and I collected for a short time in the Cordillera Volcánica and on the Mexican Plateau. In early April we moved into the Tepalcatepec Valley, where we collected intensively between Churumuco and Tepalcatepec. In May we collected on the Pacific coast between Boca de Apiza and La Placita. In July and August, 1956, accompanied by Richard E. Etheridge, we returned to Michoacán and again collected on the Mexican Plateau and in the Cordillera Volcánica, before moving into the Tepalcatepec Valley. In an attempt to fill in gaps in the known distributions of many species and to sample the fauna in some previously uncollected areas, I returned to Michoacán in June, 1958. Accompanied by Jerome B. Tulecke and John Wellman, I collected on the Mexican Plateau in the northwestern part of the state, on the southern slopes of the Cordillera Volcánica, and in the Tepalcatepec Valley. Most of our time was spent in the Sierra de Coalcomán, where we collected at Aguililla, Artega, and Dos Aguas. In 1960 two days were spent in Michoacán; a small collection was made in the eastern part of the Cordillera Volcánica. With the exception of the specimens collected in 1960, which are at the Museum of Natural History at the University of Kansas, the specimens that I have collected in Michoacán are in the Museum of Zoology at the University of Michigan.

NATURAL LANDSCAPE

A proper understanding of the geographical distribution of animals in a given region is possible only after a thorough acquaintance with the geography of the region. Likewise, in order to gain a knowledge of the ecological distribution and relationships of the components of the fauna, it is necessary to study the animals in their natural environments. In order to give the reader a picture of the physical features and the major animal habitats within the state of Michoacán, the following brief description is offered. Each of these facets mentioned below will be elaborated in detail in my final report on the herpetofauna of Michoacán.

PHYSIOGRAPHY

The state of Michoacán comprises an area of 60,093 square kilometers (Vivó, 1953). Within this area the rugged terrain has a total relief of nearly 4000 meters. There have been several attempts to classify the physiographic provinces of México; the classification used here is a slight modification of the scheme proposed by Tamayo (1949). I have tried to keep the system as simple as possible, but still useful in discussing the distribution of animals living in the region. For general purposes the state of Michoacán can be divided into lowlands and highlands as follows:

LOWLANDS
    Pacific Coastal Plain
    Balsas-Tepalcatepec Basin

HIGHLANDS
    Mexican Plateau
    Cordillera Volcánica
    Sierra de Coalcomán

Although the lowlands in the state are continuous, they are only narrowly connected and thus form two distinct physiographic and biotic areas. The Pacific Coastal Plain in Michoacán extends for a distance of about 200 kilometers (airline) from the Río Coahuayana to the Río Balsas. The coastal plain is broad between the Río Coahuayana and San Juan de Lima, and between Las Peñas and the Río Balsas, where the hills rise some 12 kilometers inland from the sea. Between San Juan de Lima and Las Peñas the mountains extend to the sea; in this region rocky promontories form precipitous cliffs dropping into the sea. Between the promontories are small sandy or rocky beaches.

Lying to the north of the Sierra de Coalcomán and the Sierra del Sur, but south of the Cordillera Volcánica, is a broad structural depression, the Balsas-Tepalcatepec Basin. The western part of this basin, which separates the Sierra de Coalcomán from the Cordillera Volcánica, is the valley of the Río Tepalcatepec, a major tributary of the Río Balsas. The eastern part of the basin is the valley of the Río Balsas. From the point of junction of the two rivers, the Río Balsas flows southward through a narrow gorge, which separates the Sierra de Coalcomán from the Sierra del Sur, to the Pacific Ocean. In Michoacán the floor of the Balsas-Tepalcatepec Basin varies from 200 to 700 meters above sea level.

The central part of México is a vast table-land, the Mexican Plateau, the southern part of which extends into northern Michoacán. In this region the terrain is rolling and varies from 1500 to 1900 meters above sea level. Many small mountain ranges rise from the plateau and break the continuity of the rolling table-land. Located on the southern part of the Mexican Plateau in Michoacán are several lakes, the largest of which are Lago de Chapala, Lago de Cuitzeo, and Lago de Pátzcuaro.

Bordering the southern edge of the Mexican Plateau is a nearly unbroken chain of volcanos, the Cordillera Volcánica. The highest peaks in Michoacán, Cerro San Andrés (3930 meters) and Cerro de Tancítaro (3870 meters), are in this range. Parts of the Cordillera Volcánica in Michoacán are known by separate names; these are, from west to east: Sierra de los Tarascos, Sierra de Ozumatlán, and Serranía de Ucareo.

Lying between the Tepalcatepec Valley and the Pacific Ocean, and east of the Río Coahuayana and west of the Río Balsas, is an isolated highland mass, the Sierra de Coalcomán. This mountain range rises to elevations of slightly more than 3000 meters. It has a length of about 200 kilometers and a width of about 80 kilometers. Except for a relatively low connection with the Cordillera Volcánica, the Sierra de Coalcomán is isolated from other mountain ranges in southwestern México.

CLIMATE

The climates in Michoacán vary from tropical in the lowlands to cool temperate at high elevations in the Sierra de Coalcomán and Cordillera Volcánica. The highest temperatures are known in the Balsas-Tepalcatepec Basin, where at Churumuco the mean annual temperature is 29.3° C. and the range of monthly means is 3.5° C. (Contreras, 1942). Frosts occur sporadically on the Mexican Plateau, and in the winter snow falls on the highest mountains.

Precipitation varies geographically and seasonally. Most of the rain falls between June and October. In the Balsas-Tepalcatepec Basin rainfall in the rest of the year is negligible. The annual average rainfall at Coahuayana on the Pacific Coastal Plain is 871 mm. (Guzmán-Rivas, 1957:52). In the Balsas-Tepalcatepec Basin rainfall seldom exceeds 800 mm. per year. In the mountains precipitation is heavier and somewhat more evenly distributed throughout the year, but still definitely cyclic. For example, Uruapan (elevation, 1500 meters) receives an average annual rainfall of 1674 mm. (Contreras, 1942). The prevailing winds are from the Pacific Ocean. The southern (windward) slopes of the Sierra de Coalcomán probably receive more rain than any other part of the state. The Balsas-Tepalcatepec Basin lies in a rain shadow of the Sierra de Coalcomán, and the Mexican Plateau lies in a somewhat less drastic rain shadow of the Cordillera Volcánica; these are the driest regions in the state.

VEGETATION AND ANIMAL HABITATS

For the purposes of this report I have adopted the classification of types of vegetation that seem to me most significant in terms of ecological distribution of reptiles and amphibians in Michoacán. These types are as follows:

TEMPERATE (1000-4000 meters)
    Fir Forest (2400-4000 meters)
    Pine-oak Forest (1000-4000 meters)
    Mesquite-grassland (1500-2100 meters)

TROPICAL (0-1000 meters)
    Arid Tropical Scrub Forest (0-1000 meters)
    Tropical Semi-deciduous Forest (150-600 meters)

The vegetation of the Pacific Coastal Plain and the Balsas-Tepalcatepec Basin consists of arid tropical scrub forest, composed of deciduous trees, which in many places are stunted and widely spaced. In the dry season there is little cover provided by this forest. In the rainy season there is a sparse growth of grasses and some shade provided by the small leaves of the thorny trees.

In Michoacán the rainfall is heaviest on the southern slopes of the Sierra de Coalcomán and somewhat less so on the southwestern slopes of the Cordillera Volcánica. At these relatively low elevations (150 to 600 meters) there is tropical semi-deciduous forest, characterized by relatively dense shade throughout the year and by a leaf mulch on the ground. This type of forest forms the gallery forest along the larger streams in the Balsas-Tepalcatepec Basin and on the Pacific Coastal Plain.

Rainfall also is heavy on the high mountain ridges, where temperatures are low. On these ridges, fir forest, often mixed with pine and oaks, is found. This habitat is characterized by a cool, moist climate, many rotting logs, and a moist ground cover of leaves and needles.

Most of the mountains are covered with pine-oak forest, which in most places is decidedly subhumid, but where this forest occurs on the windward sides of high ridges, it sometimes is noticeably humid. In this forest the important animal habitats include the needle- and leaf-litter, and in some areas, bromeliads.

The rolling terrain of the Mexican Plateau supports cacti, small leguminous trees, and grasses. Like the arid tropical scrub forest, this type of vegetation, the Mesquite-grassland association, is deciduous and thus provides little shelter in the dry season. Unlike the areas in which arid tropical scrub forest is developed, the Mesquite-grassland is found in areas having warm days and cool nights.

GEOGRAPHY OF THE HERPETOFAUNA

Although the main part of my final report on the herpetofauna of Michoacán will deal with the geographical and ecological patterns of distribution of the herpetofauna, a brief summary of the faunal assemblages is presented here.

In Michoacán there are two major faunal assemblages, one in the lowlands, and one in the highlands. A large number of the species inhabiting the lowlands are wide-ranging species, such as *Bufo marinus*, *Iguana iguana*, and *Boa constrictor*. Sixty-three species are known to occur on the Pacific Coastal Plain; 41 of these, together with 36 others occur in the Balsas-Tepalcatepec Basin, a physiographic region to which several species of reptiles are endemic; for example, *Enyaliosaurus clarki*, *Urosaurus gadowi*, *Cnemidophorus calidipes*, and *Eumeces altamirani*.

Generally speaking, the members of the highland faunal assemblage have more restricted geographic ranges. The major exceptions are those species that are widely distributed on the Mexican Plateau, such as: *Bufo compactilis*, *Sceloporus torquatus*, and *Salvadora bairdi*. In the montane habitats of the Cordillera Volcánica, 45 species of amphibians and reptiles are known; 34 species have been found in the Sierra de Coalcomán. Fourteen species are known to occur in both ranges. Several species are known only from the Cordillera Volcánica and adjacent highlands, and three species are endemic to the Sierra de Coalcomán.

ANNOTATED LIST OF SPECIES

In the following pages the 176 species and subspecies of amphibians and reptiles known to occur in the state of Michoacán are discussed in relation to their variation, life histories, ecology, and distribution in the state. Data have been gathered from 9676 specimens. I have not prolonged the accounts of species with information that has been presented elsewhere. Consequently, the length and completeness of the accounts are variable. I have given only the information that I consider a worthwhile contribution to our knowledge of the particular species.

The synonymies given at the beginning of each account include the first use of the trivial name by the original author, the first usage of the combination that I am using, and, if the circumstances make it necessary, additional names or combinations that have been proposed since the publication of the checklists of Mexican amphibians and reptiles by Smith and Taylor (1945, 1948, and 1950b). References cited only in the synonymies are not listed in the Literature Cited. Preceding the discussion of each species is an alphabetical list of the localities in Michoacán from which specimens have been examined. The listing of a locality means that one or more specimens, as indicated, has been examined from that locality. Only for those specimens especially mentioned in the text are catalogue numbers given. Abbreviations for the various museums and scientific collections are, as follows:

AMNH	American Museum of Natural History
ANSP	Academy of Natural Sciences of Philadelphia
BMNH	British Museum (Natural History)
CNHM	Chicago Natural History Museum
EHT-HMS	Edward H. Taylor-Hobart M. Smith collection
JRD	James R. Dixon collection, College Station, Texas
KU	University of Kansas Museum of Natural History
MCZ	Museum of Comparative Zoology
MVZ	Museum of Vertebrate Zoology
NMW	Naturhistorisches Museum Wien
SU	Stanford University Museum of Natural History
TCWC	Texas Cooperative Wildlife Collection
UIMNH	University of Illinois Museum of Natural History
UMMZ	University of Michigan Museum of Zoology

USNM   United States National Museum

UTNHC   University of Texas Natural History Collection

Throughout the accounts of the species all measurements are given in millimeters; if the range of variation is given, the mean follows in parentheses.

AMPHIBIA

CAUDATA

AMBYSTOMA AMBLYCEPHALUM TAYLOR

Ambystoma amblycephala Taylor, Univ. Kansas Sci. Bull., 26: 420, November 27, 1940.—Fifteen kilometers west of Morelia, Michoacán, México.

Fifteen km. W of Morelia (19); 11 km. SSE of Opopeo (12); 8 km. S of Pátzcuaro; 24 km. S of Pátzcuaro (2); Quiroga (20); Tacícuaro (167).

Taylor and Smith (1945:530) presented data on 137 specimens collected at Tacícuaro on October 1, 1939; these are all larvae and metamorphosing individuals. Aside from these, the largest larva examined (UMMZ 104962 from 15 km. W of Morelia) has a snout-vent length of 70.0 mm. and a tail length of 53.5 mm. The larvae are pale pinkish tan above and somewhat paler below; there is a lateral row of cream colored spots. The tail-fin, which is deepest at mid-length, extends to the back of the head and is flecked with brown. In small larvae the outer edge of the tail-fin is dark brown. The eyes are large. Two small metamorphosed specimens (UMMZ 98967) from 24 kilometers south of Pátzcuaro are tentatively referred to this species. These specimens have body lengths of 49.0 and 45.0 mm. and tail lengths of 36.0 and 31.5 mm., respectively. They have 17-17 and 16-15 vomerine teeth arranged in a broad arch behind the choanae, 10 costal grooves, and 7 intercostal spaces between adpressed toes. The dorsal color is uniform brown; that of the venter is a dusty cream.

Larvae were collected from shallow ponds near Quiroga and 15 kilometers west of Morelia; metamorphosed individuals were taken from beneath logs in pine and fir forests at elevations from 2300 to 2800 meters.

AMBYSTOMA DUMERILI DUMERILI (DUGÈS)

Siredon Dumerili Dugès, La Naturaleza, 1:241, 1870—Lago de Pátzcuaro, Michoacán, México.

Bathysiredon dumerilii, Dunn, Notulae Naturae, 36:1, November 9, 1939.

Bathysiredon dumerilii dumerilii, Maldonado-Koerdell, Mem. y Rev. Acad. Nac. Cien., 56:199, 1948.

Ambystoma (*Bathysiredon*) *dumerili*, Tihen, Bull. Florida State Mus., 3:3, June 20, 1958.

Lago de Pátzcuaro (22);? Morelia.

For many years this unusual salamander was known from only a few specimens mostly collected in the last century; Smith and Taylor (1948:7) stated: "It is presumed that this species is extinct owing to the introduction of exotic game and food fishes." In 1951 and in 1955 I had been told that *axolotls* were sold in the market at Pátzcuaro; nevertheless, none was found on my visits there. In 1956 Charles M. Bogert obtained several large specimens at the market in Pátzcuaro. These establish the continued existence of the salamander in Lago de Pátzcuaro. On January 27, 1955, R. W. Dickerman procured a specimen (KU 41573) in the market at Morelia. Since fish are brought to Morelia from Lago de Pátzcuaro, the specimen probably was from that lake. Nevertheless, the species may occur in other permanent bodies of water in Michoacán. Maldonado-Koerdell (1948) described *Bathysiredon dumerili queretarensis* from San Juan del Río, Queretaro. This locality is about 200 airline kilometers northeast of Lago de Pátzcuaro and is in the Río Moctezuma drainage.

AMBYSTOMA ORDINARIUM TAYLOR

Ambystoma ordinaria Taylor, Univ. Kansas Sci. Bull., 26:422, November 27, 1940.—Four miles west of El Mirador, near Puerto Hondo, Michoacán, México.

Axolotl (56); Cerro San Andrés; 22 km. W of Mil Cumbres; 46 km. E of Morelia (34); 8 km. SE of Opopeo (5); Puerto de Garnica (8); Puerto Hondo (41); San Gregorio (16); San José de la Cumbre (20).

Of 16 specimens (KU 51520-35) collected on June 18, 1955, near San Gregorio, 15 are adult females with swollen cloacae and minute ovarian eggs. Possibly these specimens had just recently deposited their mature eggs. In preservative the specimens are black above and dull creamy gray below. Measurements for the 15 females are: snout-vent length, 80.0-102.0 (92.5); tail length, 69.0-93.0 (84.2); head width, 15.8-20.5 (17.7); head length, 22.8-26.6 (24.4). A larval specimen with small gills has a snout-vent length of 72 mm. and a tail length of 62 mm. Three specimens have 12 costal grooves; the other have 11.

Of 20 specimens from San José de la Cumbre (UMMZ 112857 and 115143), 14 are neotenic adults; the others are larvae. In life the salamanders were blackish to olive-brown above with scattered cream-colored dots on the dorsum and flanks but in preservative are dull grayish black with indistinct pale spots and dark reticulations. The belly is pale gray with indistinct dark spots. Eleven females and three males have the following measurements, respectively: snout-vent length, 76.0-90.0 (80.7), 64.0-84.0 (74.3); tail length, 70.0-81.0 (75.0), 58.0-71.0 (66.7); head width, 19.5-23.5 (20.7), 17.5-20.5 (19.3); head length, 22.0-25.0 (23.0), 20.0-22.5 (21.5). The smallest larva has a snout-vent length of 43.0 mm. and a tail length of 38.0 mm. Two individuals have 12 costal grooves; the others have 11. All of the females contained eggs, the largest of which were 1.5 mm. in diameter. The stomachs of most of the specimens were distended with oligochaets, aquatic insect larvae, and small aquatic beetles.

A series of 34 larvae (JRD 5904-37) from 46 kilometers east of Morelia are tentatively referred to this species. These specimens are olive-brown above with cream-colored spots on the flanks; the dorsal tail-fin does not extend onto the body.

This species has been found only at elevations in excess of 2400 meters in pine and fir forests. At Rancho Axolotl James A. Peters collected larvae and neotenic individuals in a rocky stream and adults from beneath rocks and logs in the forest near the stream. Neotenic individuals and larvae were found in a clear stream in pine-fir forest at an elevation of 2700 meters near San José de la Cumbre; specimens were collected there in July, 1955, and again in July, 1956. The site was visited in April, 1956, at which time the stream consisted of only a few puddles; no salamanders were found.

AMBYSTOMA TIGRINUM VELASCI DUGÈS

Ambystoma velasci Dugès, La Naturaleza, ser. 2, 1:142, 1888.— Laguna Santa Isabel, near Guadalupe Hidalgo, Distrito Federal, México.

Ambystoma tigrinum velasci, Dunn, Copeia, no. 3:157, November 14, 1940.

Pátzcuaro (5); Tacícuaro (9).

Definite specific assignment of these specimens, all larvae, cannot be made at this time. They have shovel-shaped heads and laterally compressed bodies with the dorsal tail-fin extending anteriorly to the back of the head. The eyes are small. The body is pale tan with dark mottling on the tail and flanks. The average snout-vent length for nine specimens from Tacícuaro is 61.0 mm.

The larvae from Tacícuaro (UMMZ 89255) were collected by Dyfrig Forbes in October, 1939; those from Pátzcuaro, presumably Lago de Pátzcuaro (BMNH 1914.1.28-247-8 and CNHM 948), were collected by Hans Gadow and Seth Meek in 1908.

PSEUDOEURYCEA BELLI (GRAY)

Spelerpes belli Gray, Catalogue Batrachia Gradientia British Museum, p. 46, 1850.—México. Type locality restricted to 2 miles east of Río Frío, Puebla, México, by Smith and Taylor (1950a:341).

Pseudoeurycea bellii, Taylor, Univ. Kansas Sci. Bull., 30:209, June 12, 1944.

Axolotl (2); Carapan; Cerro Tancítaro (84); Macho de Agua; 22 km. W of Mil Cumbres; Opopeo; Pátzcuaro (8); Puerto Hondo (2): San José de la Cumbre; San Juan de Parangaricutiro (42); Uruapan (5); Zacapu (4).

This salamander seems to reach its greatest abundance in Michoacán in the Sierra de los Tarascos between Pátzcuaro and Tancítaro, where it is found at elevations from 1500 to 2900 meters. It is found less commonly in the eastern part of the Cordillera Volcánica in Michoacán, where it sometimes occurs in association with *Pseudoeurycea robertsi*.

On June 22 and 23, 1955, four clutches of eggs of this species were found beneath adobe bricks and rocks on the volcanic ash that has buried the village of San Juan de Parangaricutiro. The eggs were unstalked and separate, but adherent in clumps of three or four (Pl. 2, Fig. 1). The outer membranes were covered with fine particles of ash. The ash beneath the stones where the eggs were found was only slightly moist; one clump of eggs was partially desiccated. Three complete clutches have 20, 23, and 34 eggs; one clutch of 15 eggs was being eaten by beetles (Tenebrionidae: *Eleodes* sp.). The eggs vary in size from 4.6 to 6.5 mm. and average 5.3 mm. in diameter. They are unpigmented. Surrounding the embryo is a vitelline membrane, an inner, and an outer envelope (Fig. 1). In an average-sized egg having an embryo 4 mm. in length, the diameter of the outer membrane is 5.3 mm., the inner membrane 5.0 mm., and the vitelline membrane 4.6 mm. All of the eggs contained embryos in which the limb buds were developed; in about half of these the eyes were distinctly visible.

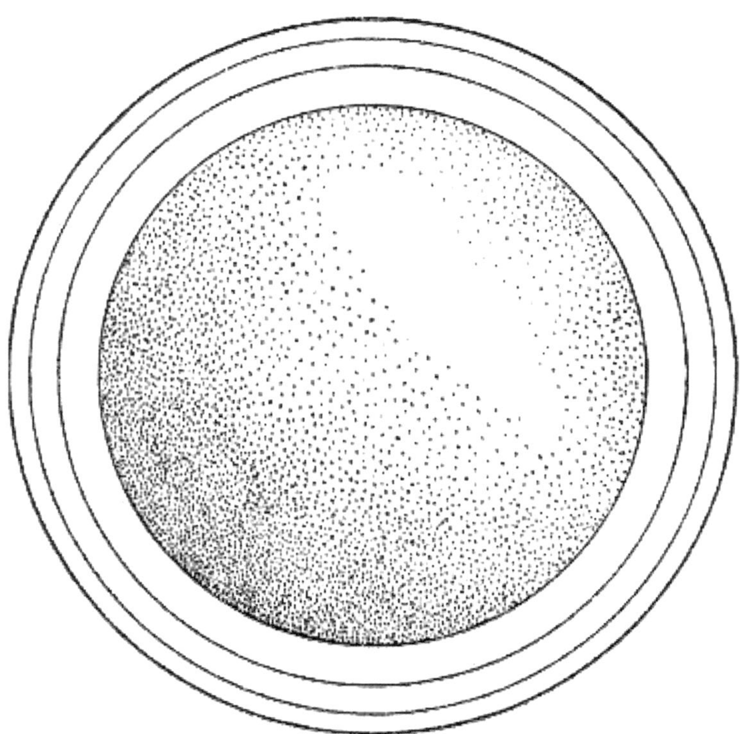

Fig. 1. Diagram of an egg of Pseudoeurycea belli from San Juan de Parangaricutiro, Michoacán. × 10.

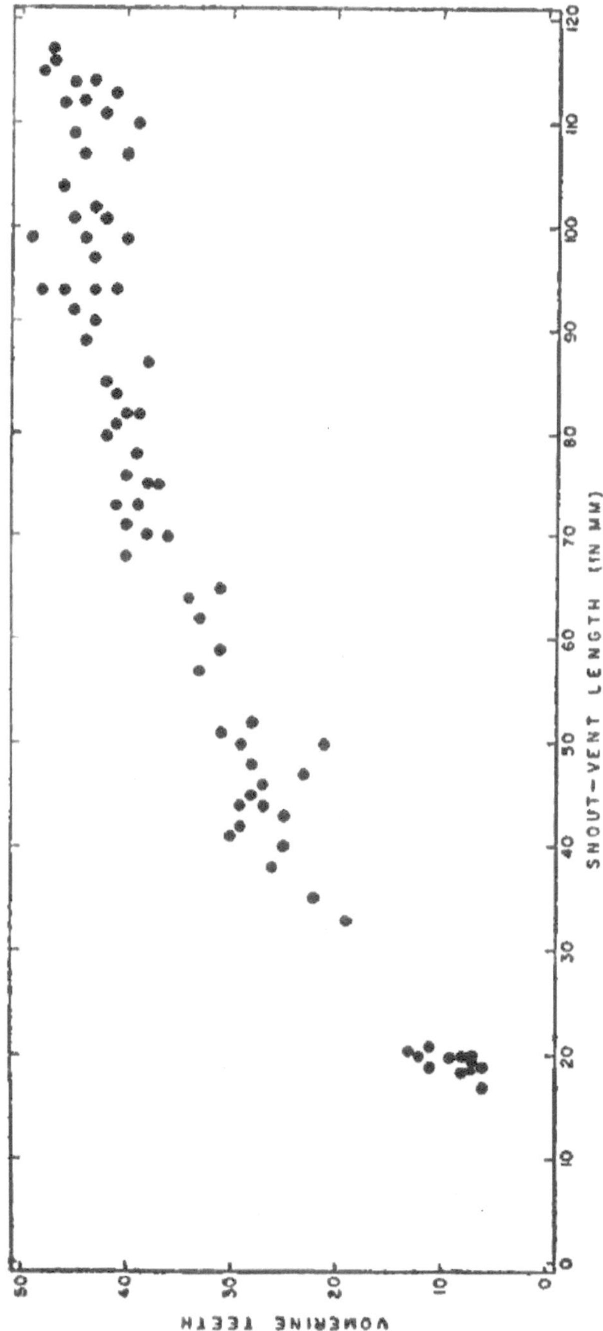

Fig. 2. Correlation between the number of vomerine teeth and snout-vent length in 79 Pseudoeurycea belli from Michoacán.

The first heavy rain of the season occurred on the night of June 22, 1955. Thus, at least sometimes, *Pseudoeurycea belli* lays its eggs before the onset of the rainy season. A female having a snout-vent length of 110 mm., collected on June 22, 1955, contained 36 ovarian eggs having diameters from 3.0 to 3.5 mm. The fact that small juveniles were collected on the same date indicates that this salamander lays eggs over a period of several weeks in late spring and early summer.

The smallest juvenile examined has a snout-vent length of 17.0 mm. and a tail length of 7.5 mm. Twelve juveniles from the vicinity of San Juan de Parangaricutiro have an average snout-vent length of 19.4 mm. and an average tail length of 9.7 mm. In juveniles the adpressed limbs either touch or overlap by one intercostal space; in adults there are two or three intercostal spaces between adpressed toes. Therefore the greatest number of intercostal spaces between adpressed limbs is found in the largest specimens. A similar relationship between adpressed limbs (= length of limbs) and snout-vent length was shown for *Plethodon richmondi* by Duellman (1954a). The number of vomerine teeth is variable; the number of teeth seems to be closely correlated with the size of the salamander (Fig. 2). A similar correlation between the number of maxillary teeth and body length was reported for *Chiropterotriton multidentatus* by Rabb (1958). In 12 juvenile *Pseudoeurycea belli* there are 6-13 (8.8) vomerine teeth, and in 11 adults having snout-vent lengths greater than 90 mm. there are 39-49 (44.0) vomerine teeth. The coloration of the juveniles resembles that of the adults (Pl. 1).

The differences between this species and *Pseudoeurycea gigantea* are minor. Taylor (1939a) distinguished *gigantea* from *belli* by the larger size, fewer intercostal spaces between adpressed limbs, more vomerine teeth, and absence of occipital spots in *gigantea*. Taylor and Smith (1945) stated that in life the spots in *gigantea* are orange instead of red as in *belli*. Five specimens of *Pseudoeurycea belli* from Michoacán, including one juvenile, lack occipital spots. In the 34 living individuals that I have seen from Michoacán the spots varied from deep red to orange. Therefore, of the characters listed by Taylor (*op. cit.*) to diagnose *Pseudoeurycea gigantea*, only the over-all larger size and smaller number of intercostal spaces between adpressed limbs (= relatively longer limbs) are useful in separating *Pseudoeurycea belli* and *gigantea*.

PSEUDOEURYCEA ROBERTSI (TAYLOR)

Oedipus robertsi Taylor, Univ. Kansas Sci. Bull., 25:287, July 10, 1939. — Nevado de Toluca, México.

Pseudoeurycea robertsi Taylor, Univ. Kansas Sci. Bull., 30:209, June 12, 1944.

Atzimba (3); Macho de Agua (9); Puerto Lengua de Vaca (14).

Previously this species has been recorded only from the type locality. In July, 1956, individuals referable to this species were found at two sites in pine-fir forest immediately to the east of Macho de Agua and in pine-oak-fir forest at Atzimba. On August 20, 1958, a series was collected in pine-fir forest at Puerto Lengua de Vaca. These localities are between 2900 and 3000 meters in the Cordillera Volcánica in eastern Michoacán.

In life the coloration of these salamanders was highly variable. The belly and undersurfaces of the tail and hind limbs were pale gray, with or without silvery white flecks; the chin was a cream-color and flecked with silvery white in some specimens. The middorsal area was brown, orange-brown, or dull grayish yellow. The flanks and lateral surfaces of the tail were black with yellowish flecks or streaks on the flanks and yellowish or orange-brown flecks on the tail. The iris was golden brown. Measurements of eight males and two females are, respectively: snout-vent length, 42.5-56.0 (49.5), 54.0-60.0 (57.0); tail length, 42.0-56.0 (48.1), 52.0-55.0 (53.5). The smallest juvenile has a snout-vent length of 28.0 mm. and a tail length of 23.0 mm. Of the 26 available specimens, six have 12 costal grooves, and the others have 11.

In comparison with 36 topotypes, the specimens from Michoacán have a less striking dorsal color pattern; none has a well-defined dorsal reddish brown area or bold reddish mottling on the tail. Furthermore, the specimens from Michoacán have paler venters than do topotypic specimens.

SALIENTIA

RHINOPHRYNUS DORSALIS DUMÉRIL AND BIBRON

Rhinophrynus dorsalis Duméril and Bibron, Erpétologie générale, vol. 8:758, 1841.—Veracruz, Veracruz, México.

Mouth of the Río Balsas (10).

These specimens (BMNH 1914.1.28.181-90) were collected by Gadow in 1908 and reported by him (1930:72): "Whilst this very sluggish termite-eating toad is common enough in the sweltering hot country of the state of Vera Cruz, up to an elevation of 1500 feet, it was unknown on the west side of the Isthmus until I found it in great numbers near the mouth of the Balsas River, in and near fresh-water pools, where it attracted attention by its loud peculiar voice during the pairing season in the month of July." Subsequently, Peters (1954:3) verified the identification of these specimens. Although torrential rains fell during the week in July, 1955, that I spent at Playa Azul near the mouth of the Río Balsas, the distinctive voice of *Rhinophrynus* was not heard. Elsewhere on the Pacific coast of México adult *Rhinophrynus* have been reported only from Tehuantepec and a few localities on the coastal lowlands of Chiapas. Taylor (1942b:37) found on the coast of Guerrero a tadpole that was referred to the genus *Rhinophrynus* by Orton (1943). In the summer of 1960 adults of *Rhinophrynus* were collected near Acapulco, Guerrero (Fouquette, *in litt.*). These recent collections verify the existence of the species along the Pacific lowlands of México at least as far north as Michoacán.

SCAPHIOPUS HAMMONDI MULTIPLICATUS COPE

Scaphiopus multiplicatus Cope, Proc. Acad. Nat. Sci. Philadelphia, 15:52, June 8, 1863.—Valley of México.

Scaphiopus hammondi multiplicatus, Kellogg, Bull. U. S. Natl.

Mus., 160:22, March 31, 1932.

Angahuan (5); Cuitzeo (4); Cuseño Station (2); Jiquilpan (9); Morelia (7); Pátzcuaro (3); Quiroga; Tarécuaro; Uruapan (24); Zacapu.

This small toad has been found at elevations between 1500 and 2500 meters on the Mexican Plateau and associated mountain ranges; it occurs in mesquite-grassland and in pine forests. Calling males and females laden with eggs have been collected in the rainy season in the months of July and August. The call is a medium-pitched snore. In living individuals the dorsal ground color varies from pale brown to gray with dark brown or olive-brown markings. In many individuals the tips of the small dorsal pustules are red.

BUFO COCCIFER COPE

Bufo coccifer Cope, Proc. Acad. Nat. Sci. Philadelphia, 18:130, 1866—Arriba, Costa Rica.

Apatzingán (27); Lombardia; Nueva Italia (5).

In life the dorsal color pattern consists of a yellowish tan ground color with dark brown spots; the middorsal stripe is deep yellow or cream color. The venter is a dusty cream color, and the iris is pale gold. Males have dark brown horny nuptial tuberosities on the thumb. The following measurements are of 21 males and four females, respectively: snout-vent length, 43.5-51.7 (48.1), 55.6-62.6 (59.1); tibia length, 16.6-18.8 (17.6), 18.8-20.3 (19.3); head width, 16.7-19.7 (18.4), 20.6-22.2 (21.4); head length, 13.8-16.6 (14.8), 16.5-18.2 (17.3).

The specimens from the Tepalcatepec Valley differ slightly from specimens from southeastern México and Central America. Those from Michoacán have low and narrow cranial crests; in about one-half of the specimens the occipital crest exists only as a row of tubercles, and in some the postorbital and suborbital crests are barely discernible. Specimens from the southern part of the range, Costa Rica and Nicaragua, have much higher and thicker cranial crests; in these the occipital crest is well defined and extends posteriorly to a point back of the anterior edge of the parotid gland; the postorbital and suborbital crests are well marked. Of 48 specimens from Esquipulas, Guatemala, all have high crests, but these are not so well developed as in ten specimens from Matagalpa, Nicaragua, and three from various localities in Costa Rica. Six specimens from Tehuantepec, Oaxaca, have cranial crests that are lower than those in specimens from Guatemala. In three of the specimens from Tehuantepec the occipital crests are reduced to a series of tubercles. Of six specimens from Agua del Obispo, Guerrero, four have poorly developed occipital crests. These observations suggest the presence of a cline in the development of the cranial crests; specimens have higher crests in the southern part of the range than in the northern part.

In México *Bufo coccifer* has been collected only in semi-xeric habitats, but to the south, from Guatemala to Costa Rica, it has been found in more upland and humid habitats. Southern specimens are darker than those from the north, a possible correlation with the differences in habitat.

These toads probably range throughout the Tepalcatepec Valley, but they are unknown from the coast of Michoacán. Breeding choruses were found after heavy rains on June 24, 1955, and on August 2, 1956. The first was in a muddy ditch; the second was in a flooded grassy field. The call is a high-pitched, but not loud, "whirrr." Males were calling from the edge of the water or from clumps of grass in the water. Clasping pairs were in the water; amplexus is axillary.

BUFO COMPACTILIS COMPACTILIS WIEGMANN

Bufo compactilis Wiegmann, Isis von Oken, 26:661, 1833.— México. Type locality restricted to Xochimilco, Distrito Federal, México, by Smith and Taylor (1950a:330).

Bufo compactilis compactilis, Smith, Herpetologica, 4:7, September 17, 1947.

Cuitzeo (2); Emiliano Zapata (20); Jiquilpan (5); La Palma (5); Morelia; Tupátaro.

The southwestern terminus of the range of this species is on the Mexican Plateau in Michoacán. All specimens from the state have spotted venters. In living toads the dorsal ground color was gray or grayish tan with olive green spots. The vocal sac was brownish gray; the iris was a bright golden color.

On June 11, 1958, many individuals were calling from shallow water in a flooded field at Emiliano Zapata. The call is a slow trill, in which the individual notes are discernible.

BUFO MARINUS (LINNAEUS)

Rana marina Linnaeus, Systema naturae, ed. 10, 1:211, 1758.— America.

Bufo horribilis Wiegmann, Isis von Oken, 26:654, 1833.—Misantla and Veracruz, Veracruz, México. Taylor and Smith, Proc. U. S. Natl. Mus., 95:551, January 30, 1945.

Bufo angustipes Taylor and Smith, Proc. U. S. Natl. Mus., 95:553, January 30, 1945.—La Esperanza, Chiapas, México.

Aguililla; Apatzingán (3); Barranca de Bejuco; Capirio; Charapendo; Chichihuas; Coahuayana (2); Coalcomán (7); Cofradía (2); 25 km. S of Cuatro Caminos; El Sabino (10); Huahua, La Playa (13); Ojos de Agua de San Telmo; Ostula; Playa Azul (2); Pómaro (2).

This large toad is characteristically found in areas supporting tropical scrub forest to elevations of about 1000 meters. The species is much more abundant than the numbers listed above suggest. In the dry season individuals have been observed in patios, along streams, and by irrigation ditches. In the rainy season the loud, rattling call of the males is heard at night throughout the Tepalcatepec Valley and the coastal lowlands.

Taylor and Smith (1945:552) revived Wiegmann's *Bufo horribilis* for the large toads of México that are here referred to *B. marinus*. Their action was based upon the supposition that the "species *marinus*" is composite. Although probably true, this supposition has yet to be proved. Until the large, and apparently related, species of *Bufo* inhabiting tropical America have been studied systematically as a unit, the recognition of segments of the population as either species or subspecies is meaningless. Taylor and Smith (op. cit.:553) based the description of a new species, *Bufo angustipes*, on one rather emaciated, formalin-hardened female from La Esperanza, Chiapas. The type (USNM 116513), when compared with numerous specimens of *Bufo marinus* from throughout the range of the species in México and northern Central America, displays no combination of characters to set it off from the others. Therefore, I suggest that *Bufo horribilis* Wiegmann and *Bufo angustipes* Taylor and Smith be placed in the synonymy of *Bufo marinus* (Linnaeus) until future systematic study of the genus and this species in particular establishes the existence of recognizable taxa.

BUFO MARMOREUS WIEGMANN

Bufo marmoreus Wiegmann, Isis von Oken, 26:66, 1833.—Veracruz, Veracruz, México.

Barranca de Bejuco; Coahuayana (11); El Diezmo (2); La Placita (9); La Orilla (12); Motín del Oro; Ostula (9); Playa Azul (5); Pómaro (15); Salitre de Estopilas; San Pedro Naranjestila.

In Michoacán this species is confined to elevations of less than 1000 meters on the coast and foothills of the Sierra de Coalcomán. In this region in the months of June and July, breeding congregations have been found in temporary pools and along streams.

Smith and Taylor (1948:39), in their key to the Mexican species of *Bufo*, placed emphasis on the nature of the supraorbital and postorbital crests (whether they form a curve or a sharp angle) in distinguishing *Bufo marmoreus* from *Bufo perplexus*. In the original description of *perplexus*, Taylor (1943a:347) characterized the species as follows: supraorbital and postorbital crests forming a sharp angle, instead of a curve as in *marmoreus*; supratympanic crest smaller than in *marmoreus*; diagonal lateral stripe lacking in females; concentration of dorsal tubercles as found in *marmoreus* lacking in males. The discovery of specimens in which the crests form a curve and others in which the crests form an angle in both the Tepalcatepec Valley and in the coastal lowlands prompted an investigation of these characters and others throughout the ranges of the species. An examination of 410 specimens has resulted in the following conclusions.

TABLE 1.—VARIATION IN THE SHAPE OF THE SUPRAORBITAL AND POSTORBITAL CRANIAL CRESTS IN BUFO MARMOREUS AND B. PERPLEXUS.

Locality	N	Curved	Intermediate	Angular
Tepalcatepec Valley	50	10 (20.0%)	17 (34.0%)	23 (46.0%)
Morelos	12	2 (16.6%)	5 (41.7%)	5 (41.7%)
Izúcar, Puebla	4	2 (50.0%)	0 (0.0%)	2 (50.0%)
Southern Sinaloa	1	1(100.0%)	0 (0.0%)	0 (0.0%)
Puerto Vallarta, Jalisco	2	2(100.0%)	0 (0.0%)	0 (0.0%)
Colima	45	25 (55.0%)	18 (40.0%)	2 (5.0%)
Coast of Michoacán	55	35 (63.6%)	17 (30.9%)	3 (5.5%)
Acapulco, Guerrero	7	7(100.0%)	0 (0.0%)	0 (0.0%)
Chilpancingo, Guerrero	10	1 (10.0%)	4 (40.0%)	5 (50.0%)
Pochutla, Oaxaca	13	6 (46.2%)	6 (46.2%)	1 (7.6%)
Tehuantepec, Oaxaca	177	81 (45.8%)	67 (37.8%)	29 (16.4%)
Tonolá, Chiapas	1	0 (0.0%)	0 (0.0%)	1(100.0%)
Veracruz	33	26 (78.8%)	6 (18.2%)	1 (3.0%)
Total	410	198 (48.3%)	140 (34.2%)	72 (17.5%)

1. Although the highest percentage of individuals having the supraorbital and postorbital crests forming a sharp angle is from localities in the Balsas-Tepalcatepec Basin, numerous individuals from throughout the range of *marmoreus* have the crests forming an angle (Table 1).

2. In all samples of ten or more specimens, some toads have the supraorbital and postorbital crests forming a sharp angle, some have the crests forming a curve, and some have an intermediate condition.

3. The relative size of the supratympanic crest is highly variable in all samples examined.

4. A distinct, pale-colored, diagonal lateral stripe is found in females only from localities outside of the Balsas-Tepalcatepec Basin; females from the basin have a spotted dorsum.

5. Males from the Balsas-Tepalcatepec Basin usually have a broad middorsal line that is yellow or pale tan; those from outside the basin have either a narrow middorsal line or none.

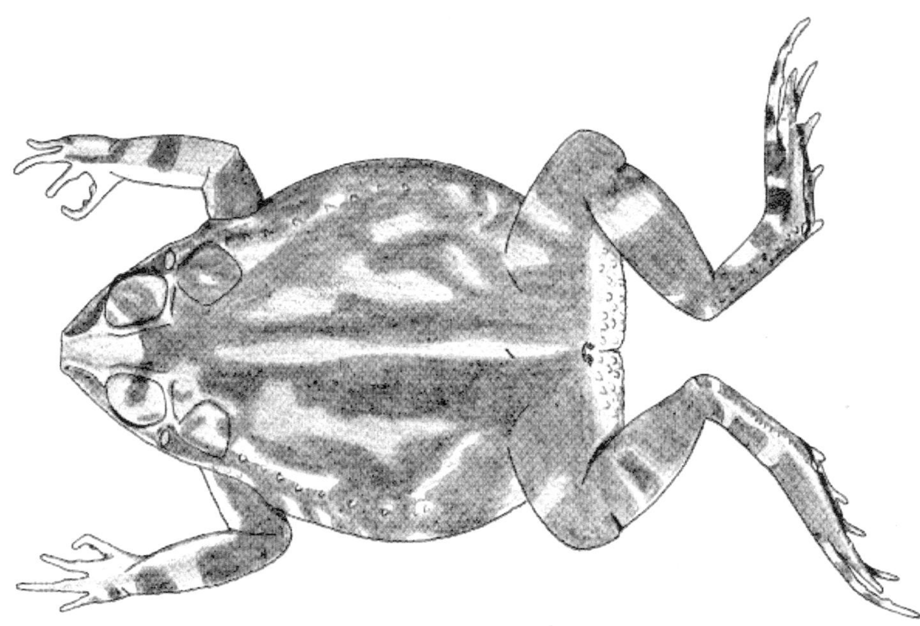

Fig. 3. Adult male of Bufo perplexus from Apatzingán, Michoacán. × 1.5.

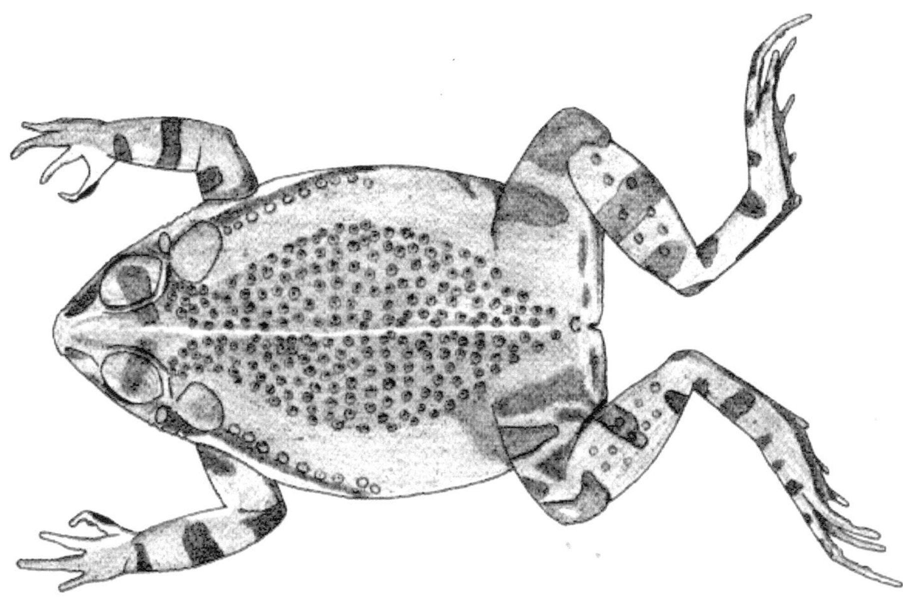

Fig. 4. Adult male of Bufo marmoreus from Pómaro, Michoacán. × 1.5.

PLATE 1

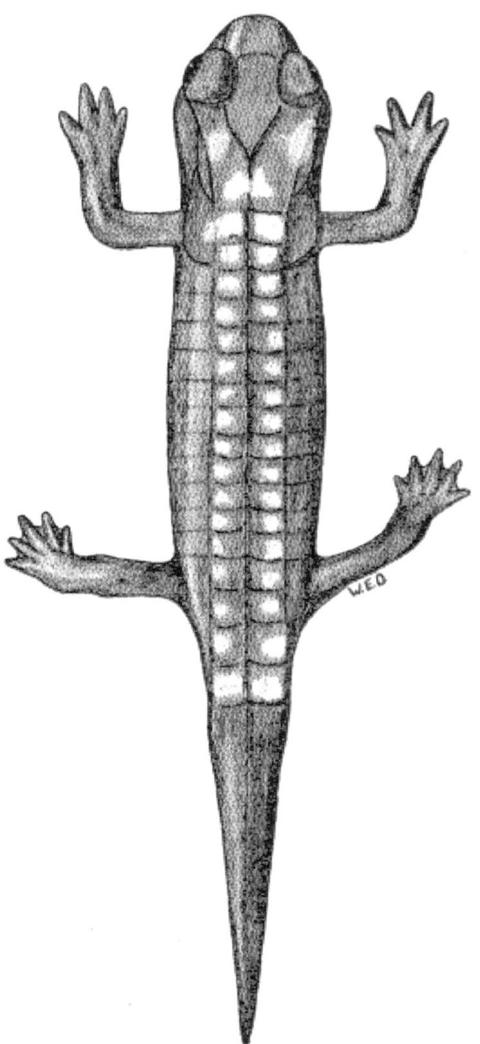

Hatchling of Pseudoeurycea belli from San Juan de Parangaricutiro, Michoacán. × 8.

PLATE 2

Fig. 1. Nest and eggs of Pseudoeurycea belli beneath a rock at San Juan de Parangaricutiro. Approx. natural size.

Fig. 2. Multiple egg clutches of Phyllomedusa dacnicolor from Coalcomán, Michoacán. 1/3 ×.

PLATE 3

Fig. 1. Adult male of Tomodactylus angustidigitorum from Paracho, Michoacán. × 4.

Fig. 2. Adult male of Tomodactylus fuscus from Los Cantiles, Michoacán. ×4.

PLATE 4

Fig. 1. Adult male of Tomodactylus nitidus nitidus from Tuxpan, Michoacán. ×.

Fig. 2. Adult male of Tomodactylus nitidus orarius from Tecolapa, Colima. × 4.

PLATE 5

Fig. 1. Adult male of Tomodactylus nitidus petersi from Apatzingán, Michoacán. × 4.

Fig. 2. Adult male of Tomodactylus rufescens from Dos Aguas, Michoacán. × 4.

PLATE 6

Fig. 1. Adult male of Hypopachus caprimimus from Tuxpan, Michoacán. ×
2-1/2.

Fig. 2. Adult male of Hypopachus oxyrrhinus ovis from Tangamandapio, Michoacán. × 3.

6. Males from the Balsas-Tepalcatepec Basin have low, scattered dorsal tubercles (Fig. 3); males from outside the basin have a concentration of tubercles in a broad band on the back (Fig. 4).

Therefore the nature of the cranial crests is of little value in separating two populations, but the color pattern of the females and the nature of the dorsal tubercles of the males do show distinct differences. Furthermore, certain differences in size and proportion are evident; *Bufo marmoreus* is a slightly larger toad and has a relatively longer tibia and longer head than *perplexus* (Table 2).

TABLE 2.—COMPARISON OF CERTAIN MEASUREMENTS AND PROPORTIONS IN BUFO MARMOREUS AND B. PERPLEXUS. (MEANS ARE GIVEN IN PARENTHESES BELOW THE RANGES.)

Species	Sex	N	Snout-vent length	Tibia length Snout-vent length	Head length Snout-vent length
B. marmoreus	♂	15	61.5-72.5	35.9-41.6	28.3-33.3
			(65.2)	(39.0)	(31.6)
B. perplexus	♂	20	50.0-59.0	33.7-38.1	26.4-31.1
			(54.9)	(36.4)	(29.5)
B. marmoreus	♀	7	68.0-76.0	33.0-36.8	26.8-32.6
			(70.7)	(34.7)	(29.6)
B. perplexus	♀	6	64.1-69.8	32.4-36.9	25.1-29.0
			(66.8)	(35.5)	(27.5)

Taylor (1943a:347) described *Bufo perplexus* from Mexcala on the Río Balsas in Guerrero. Among the many paratypes are specimens from Tonolá, Chiapas, and Tehuantepec, Oaxaca. These apparently were referred to *perplexus* solely on the nature of the cranial crests. All of the specimens examined during the course of the present study from the lowlands of Veracruz and from the Pacific lowlands from Sinaloa southward to Chiapas are referable to *Bufo marmoreus*; those from the Balsas-Tepalcatepec Basin are referable to *Bufo perplexus*, as defined above. Ten specimens from Chilpancingo, Guerrero (UMMZ 115352), do not readily fit either species. Perhaps there is gene exchange between the inland and coastal populations through the relatively low pass at Chilpancingo, at the mouth of the Río Balsas, and near the convergent headwaters of the Río Coahuayana and Río Tepalcatepec in southern Jalisco. If this can be demonstrated, then *Bufo perplexus* would have to be considered as a subspecies of *Bufo marmoreus*, instead of an allopatric species.

BUFO PERPLEXUS TAYLOR

Bufo perplexus Taylor, Univ. Kansas Sci. Bull., 29:347, October 15, 1943.—Balsas River near Mexcala, Guerrero, México.

Aguililla (2); Apatzingán (42); Buena Vista (5); Capirio (3); La Playa (25); Lombardia (6); Nueva Italia (9); Río Cancita, 14 km.

E of Apatzingán; Río Tepalcatepec, 27 km. S of Apatzingán; San Salvador (4); Tzitzio; Volcán Jorullo.

BUFO OCCIDENTALIS CAMERANO

Bufo occidentalis Camerano, Atti R. Accad. Sci. Torino, 14:887, December 31, 1878.—México. Type locality restricted to Guanajuato, Guanajuato, México, by Smith and Taylor (1950a:330). Firschein, Copeia, no. 3:220, September 15, 1950.

Bufo símus, Smith and Taylor, Bull. U. S. Natl. Mus., 194:42, 1948.

Barranca Seca (32); Cerro de Barolosa (4); Cerro Tancítaro, 3 km. E of Apo (2); Cerro Tancítaro, 19 km. E. of Apo (10); Charapendo; Coalcomán (7); Dos Aguas (4); Jacona, Jaramillo (2); Las Tecatas; Los Reyes (181); Tancítaro (10); Uruapan (3).

This toad is an inhabitant of pine and oak forests between 900 and 2400 meters. Near Charapendo on the slopes of the Sierra de los Tarascos and at Coalcomán it apparently reaches its lowest altitudinal limits. At both of these localities the pine-oak forest is replaced by arid tropical scrub forest on the lower slopes.

Twenty-four tadpoles were collected on May 3 in a quiet section of a fast stream near Barranca Seca. The tadpoles have a robust body, broadest about two-thirds the distance from the snout to the posterior edge of the body, half again as broad as deep. Eyes dorsolateral; nostrils dorsal, somewhat directed forward, and about three-fifths the distance from the tip of the snout to the eye; spiracle sinistral and lateral, located at about midbody; anus median; tail long and slender; tail-musculature extends nearly to tip of tail; depth of tail-musculature at mid-length about one-third total depth of tail; dorsal tail-fin not extending onto body (Fig. 5); average body length of ten tadpoles having small hind limb buds, 14.4 mm.; average tail length, 22.0 mm.

Mouth ventral, nearly terminal, about one-third as wide as widest part of body; anterior lip has no papillae; lower lip bordered by two rows of papillae and lateral lips by one row of papillae; beaks moderately well developed, the upper forming a broad arch and finely denticulate; tooth rows 2/3, the upper rows extending to the edge of the lips, subequal in length, and slightly longer than lower rows, which also are subequal in length; inner upper tooth row broken medially; inner lower tooth row sometimes broken (Fig. 6).

The body is black dorsally and laterally, and bluish gray ventrally; the tail musculature is brown and stippled with darker brown. The fins are transparent and stippled with brown, the stippling being most pronounced on the posterior two-thirds of the upper tail-fin.

Forty recently metamorphosed individuals average 18.9 mm. in snout-vent length.

The relationships of this toad seem to be with *Bufo bocourti* Brocchi, an inhabitant of pine and oak forests in the uplands of Chiapas and Guatemala. In *Bufo oc-*

cidentalis the tympanum usually is indistinct and sometimes completely covered, and it is absent in *bocourti*. *Bufo occidentalis* has a broader interorbital area and relatively shorter and more rounded parotid glands than *bocourti*. The tadpoles of the two species are nearly identical (see Stuart, 1943:12).

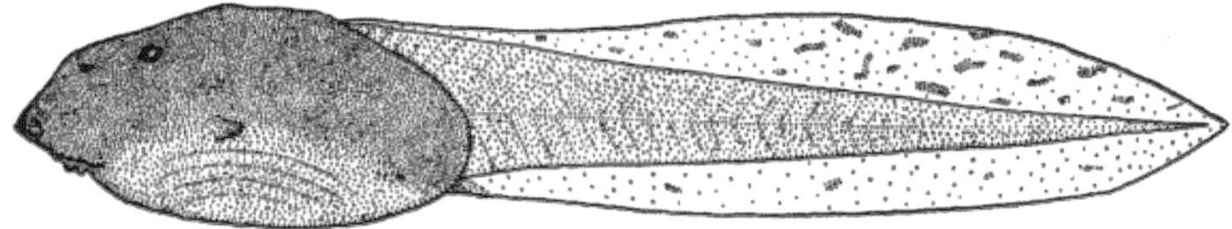

Fig. 5. Tadpole of Bufo occidentalis (UMMZ 94269) from Barranca Seca, Michoacán. × 3.

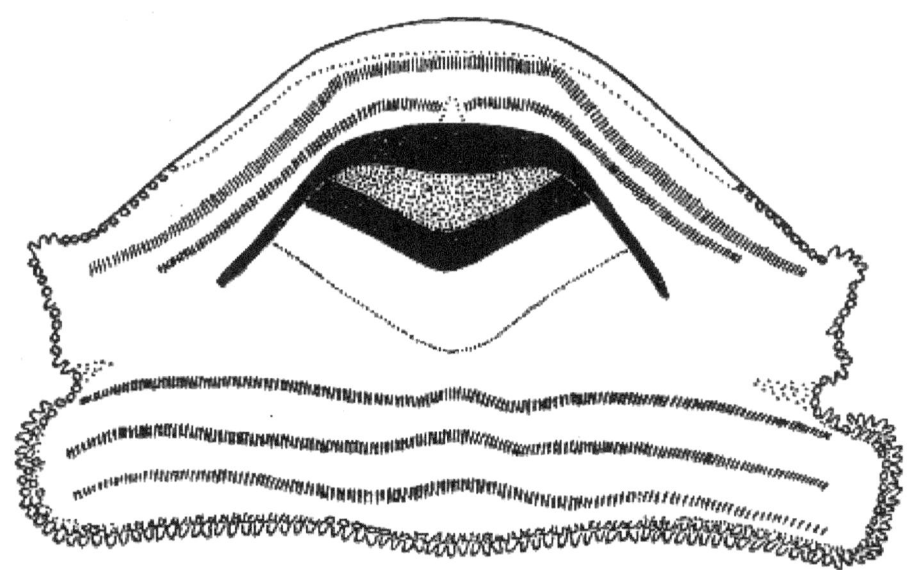

Fig. 6. Mouthparts of larval Bufo occidentalis (UMMZ 94269) from Barranca Seca, Michoacán. × 20.

LEPTODACTYLUS LABIALIS (COPE)

Cystignathus labialis Cope, Proc. Amer. Philos. Soc., 17:90, 1877.—No type locality designated; type locality restricted to Potrero Viejo, Veracruz, México, by Smith and Taylor (1950a:350).

Leptodactylus labialis, Brocchi, Mission Scientifique au Mexique et dans l'Amerique Centrale, pt. 3, sec. 2, livr. 1:20, 1881.

Apatzingán (26); Capirio (5); Cofradía (9); El Sabino (4); Lombardia; Río Tepalcatepec, 27 km. S of Apatzingán (2).

In the Tepalcatepec Valley this frog reaches the northernmost known limit of its range in western México. Although the species is abundant in the valley, it apparently is absent from the coastal lowlands. In the Tepalcatepec Valley *Leptodactylus melanonotus* seems to be more abundant than *labialis*. In the rainy season both species have been heard calling from the same ponds and flooded fields.

There are only slight differences in size between the sexes; measurements of 20 males and eight females are, respectively: snout-vent length, 32.3-39.5 (35.1), 34.1-39.2 (37.2); tibia length, 14.3-17.0 (15.4), 14.9-16.8 (15.8); head width, 11.0-13.6 (12.0), 12.2-13.2 (12.6); head length, 12.8-15.1 (13.3), 12.8-14.6 (13.7).

LEPTODACTYLUS MELANONOTUS (HALLOWELL)

Cystignathus melanonotus Hallowell, Proc. Acad. Nat. Sci. Philadelphia, 12:485, 1861.—Nicaragua. Type locality restricted to Recero, Nicaragua, by Smith and Taylor (1950a:320).

Leptodactylus melanonotus, Brocchi, Mission Scientifique au Mexique et dans l'Amerique Centrale, pt. 3, sec. 2, livr. 1:20, 1881.

Apatzingán (103); Capirio; Charapendo (7); Coahuayana; Cofradía (10); El Sabino (21); La Playa (3); Lombardia (5); Maruata; Nueva Italia (7); Ostula (9); Playa Azul (11); Río Marquez, 10 km. S of Lombardia; Río Marquez, 13 km. SE of Nueva Italia (6); Río Tepalcatepec, 27 km. S of Apatzingán.

This species is widespread in the lowlands of the state; it has been collected up to elevations of 1050 meters in the Tepalcatepec Valley. In the dry season individuals were discovered beneath rocks along streams and in damp arroyos; in the rainy season they were found wherever there was water. Males were heard calling from flooded fields, ditches, rocky streams, and small puddles. The call is a series of individual notes: "woink, woink, woink."

Adult males are noticeably smaller than females; measurements for 20 males and ten females from Apatzingán are, respectively: snout-vent length, 29.6-34.6 (32.3), 36.3-44.1 (40.8); tibia length, 12.6-15.1 (14.0), 16.5-19.0 (17.8); head width, 10.8-11.9 (11.3), 12.6-14.8 (13.7); head length, 11.2-13.2 (11.9), 13.1-14.8 (14.0). Brownish yellow ventral glands are present in some juveniles and in some adults collected in the dry season as well as in the rainy season.

LEPTODACTYLUS OCCIDENTALIS TAYLOR

Leptodactylus occidentalis Taylor, Trans. Kansas Acad. Sci., 39:349, 1937.—Tepic, Nayarit, México.

Five km. W of Tangamandapio.

On the night of June 11, 1958, this species was calling from a hyacinth-choked ditch. Although numerous individuals were heard, only one specimen was obtained. The frogs were calling from the tangled mat of hyacinths along with *Hyla eximia, Hypopachus oxyrrhinus ovis,* and *Rana pipiens.*

Taylor (1936a:352) characterized this species as follows: "The narrow head, small maximum size (38 mm. for females, 33 mm. for males), the character of the postaxillary and postfemoral glands, the narrower groups of vomerine teeth, clearly distinguish this western Mexican form from the more robust, larger *melanonotus* to the south. The call is likewise fainter and different in quality." Concerning the glands, Taylor (*loc. cit.*) remarked: "There is a possibility that the horny excrescence covering the glands may appear only during the breeding season. This character is quite as strongly marked in females as in males." Bogert and Oliver (1945:324) concluded that the population of *Leptodactylus* in northwestern México could not be distinguished from *melanonotus* in other parts of the country and thus synonymized *Leptodactylus occidentalis* with *melanonotus*. Bogert and Oliver (op. cit.: 324) stated that the extent as well as the presence or absence of ventral glands was highly variable in all samples examined by them.

Upon seeing numerous living individuals of *Leptodactylus melanonotus* from many parts of its range in México and individuals of the population of *Leptodactylus* in northwestern México (Nayarit and Sinaloa), I was immediately impressed not so much by the differences in the development of the ventral glands, but by the color of the glands. The differences in color are apparent in freshly preserved specimens. With the exception of *Leptodactylus* from northwestern México, specimens of *melanonotus* from throughout México and northern Central America have yellow or yellowish brown glands. Specimens from northwestern México have black or brownish black glands that are conspicuously darker than those found in *melanonotus*. Examination of 653 preserved specimens of *Leptodactylus melanonotus* from México and Guatemala has failed to reveal specimens with black ventral glands, like those found in specimens from northwestern México, to which the name *Leptodactylus occidentalis* has been applied. Furthermore, in *melanonotus* the glands are less distinct and more extensive than in *occidentalis*; in the latter species glands are absent from the throat and midventral area, where they often are present in *melanonotus* (Fig. 7).

In some individuals of both species collected in the dry season and in some collected in the rainy (breeding) season the glands are absent; the development of these glands, therefore, does not seem to be correlated with breeding. Likewise, the glands are present or absent in either sex, and often as not they are present in juveniles. Presence of the glands, therefore, cannot be correlated either with sexual or ontogenetic development. Since the glands are found in individuals from all parts of the range, it is unlikely that there is a correlation between the development of the glands and the environment.

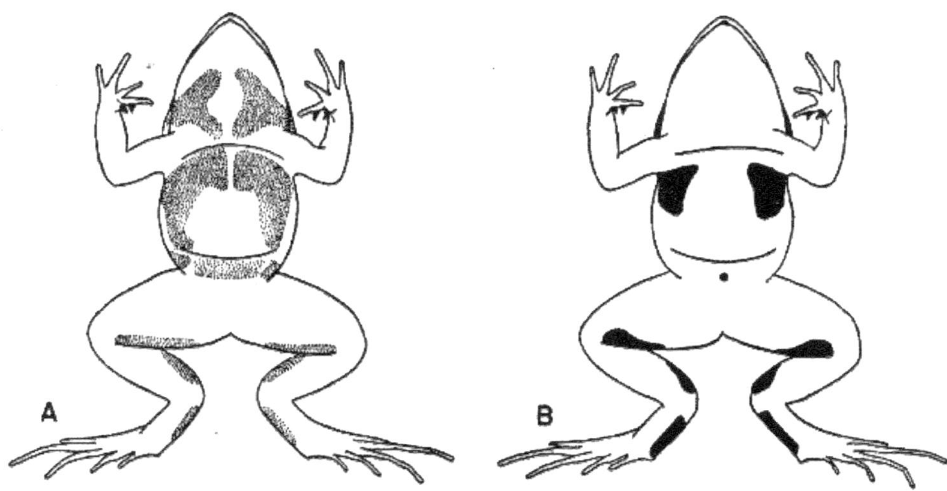

Fig. 7. Diagrammatic view of ventral surfaces of Leptodactylus melanonotus (A) and Leptodactylus occidentalis (B), showing usual position and size of glandular areas. Approx. natural size.

Aside from the differences in the ventral glands, the call is different in the two populations. The call of *Leptodactylus occidentalis* is a rather harsh "wack, wack, wack" as contrasted with the more nasal "woink, woink, woink" of *melanonotus*. Sound spectrographs are needed to analyze the differences in calls. None of the specimens of *occidentalis* examined approaches in size the largest individuals of *melanonotus*; possibly the size of the frogs is another valid character for separating the species. On the basis of the above data it is evident that the frogs in north-western México show certain characters that distinguish them from *Leptodactylus melanonotus*, as it is known throughout the rest of México. It is not known for certain that *melanonotus* and *occidentalis* are sympatric. Several series of old, poorly preserved specimens from Nayarit and Sinaloa cannot be placed in either species, for none has visible ventral glands. *Leptodactylus melanonotus* is known from Acaponeta, Nayarit (AMNH 43913-25), and the following localities in Jalisco: Barro de Navidad (UMMZ 118098), La Concepción (UMMZ 113081), La Resolana (UMMZ 102104), and Tenachitlán (UMMZ 113045-6). Records for *Leptodactylus occidentalis* are: Álamos, Sonora (AMNH 51356-65); Culiacán (AMNH 49511-9), Chele (UMMZ 110914), and Rosario (UMMZ 113062) in Sinaloa; Ixtlán del Río (UMMZ 102108), San Blas (UMMZ 112814, 112994, 110892, 115543), and Tepic (UMMZ 115544) in Nayarit; Ameca (UMMZ 102106-7) and La Cofradía on the south shore of Lago de Chapala (UMMZ 102105) in Jalisco; and Tangamandapio, Michoacán (UMMZ 119145). From these scattered records it appears that *Leptodactylus occidentalis* in the southern part of its range stays in the uplands, whereas *melanonotus* is confined to the lowlands.

MICROBATRACHYLUS HOBARTSMITHI (TAYLOR)

Eleutherodactylus hobartsmithi Taylor, Trans. Kansas Acad. Sci., 39:355, 1937.—Uruapan, Michoacán, México.

Microbatrachylus hobartsmithi Taylor, Univ. Kansas Sci. Bull., 26:501, November 27, 1940.

Cascada Tzararacua (6); 21 km. W of Ciudad Hidalgo; 29 km. E of Morelia; Puerto Hondo; San José de la Cumbre (13); Uruapan (2); Zitácuaro.

Of six specimens from Cascada Tzararacua, five are colored like typical *M. hobartsmithi*, having the anterior and posterior surfaces of the thighs and the upper arms pale pink in life and a grayish brown dorsum in preservative. The other specimen (UMMZ 94231) has in preservative a dark brown dorsolateral line on each side enclosing a pale tan area that extends from the snout to the vent. One specimen from 29 kilometers east of Morelia (UIMNH 40338) and 13 specimens from San José de la Cumbre (UMMZ 102111) do not have the prominent tarsal tubercles characteristic of *M. hobartsmithi*. Also, in these fourteen specimens the palmar tubercles are larger, and the dark anal patch more distinct, than in typical *M. hobartsmithi*. Possibly these specimens, which are from the high mountains in the eastern part of Michoacán, represent another species of *Microbatrachylus*. However, Taylor (1940d:501) reported a series of *M. hobartsmithi* from the mountains 10 miles west of Villa Victoria in the western part of the state of México.

The largest specimen from Michoacán is a gravid female (UIMNH 16104) having a snout-vent length of 23.5 mm.

Microbatrachylus hobartsmithi has been found in rocky ravines along streams in the Cordillera Volcánica and the southwestern escarpment of these mountains at elevations from 1450 to 2750 meters.

MICROBATRACHYLUS PYGMAEUS (TAYLOR)

Eleutherodactylus pygmaeus Taylor, Trans. Kansas Acad. Sci., 39:352, 1937.—1 mile north of Rodriguez Clara, Veracruz, México.

Microbatrachylus pygmaeus Taylor, Univ. Kansas Sci. Bull., 26:500, November 27, 1940.

Microbatrachylus albolabris Taylor, Univ. Kansas Sci. Bull., 26:502, November 27, 1940.—2 miles west of Córdoba, Veracruz, México.

Microbatrachylus minimus Taylor, Univ. Kansas Sci. Bull., 26:507, November 27, 1940.—Agua del Obispo, Guerrero, México.

Microbatrachylus imitator Taylor, Univ. Kansas Sci. Bull., 28:70, May 15, 1942.—La Esperanza, Chiapas, México.

Arteaga (328).

This large series (UMMZ 119247-8) was collected on June 22 and 23, 1958, before the onset of the heavy summer rains. The frogs were found in a shaded ravine at the north edge of Arteaga; they were obtained during the day, at which time they were actively moving about in the leaf litter along a small stream.

These frogs are all referred to *M. pygmaeus*, because this is the earliest name available for frogs showing the variation in characteristics displayed by this large series. The characters used by Taylor (1936a, 1940d, 1941a, and 1942b) and Smith and Taylor (1948) to distinguish the various species of *Microbatrachylus* include color pattern, relative length of the hind limb, presence and position of dorsal dermal folds or pustules, relative size of inner and outer metatarsal tubercles, and the number of palmar tubercles. All specimens from Arteaga have two palmar tubercles; the inner and outer metatarsal tubercles are subequal in size. Furthermore, aside from sexual difference, there is little variation in the relative length of the hind limbs (Table 3). However, many color patterns do exist in the series; each of these color patterns is described below.

TABLE 3.—SNOUT-VENT LENGTH EXPRESSED AS A PERCENTAGE OF TIBIA LENGTH IN ANIMALS OF SIX COLOR PATTERNS OF MICROBATRACHYLUS PYGMAEUS. (LETTERS REFER TO THE VARIANTS HAVING THE COLOR PATTERN DISCUSSED IMMEDIATELY BELOW)

Color Pattern	Sex	Number of specimens	Range of variation	Mean	Twice standard error of mean
A	♂	25	51.4-57.5	55.2	3.34
A	♀	25	49.3-54.9	51.6	3.12
B	♂	20	51.0-57.1	55.4	2.44
B	♀	21	47.3-54.9	51.2	3.52
C	♂	6	54.5-56.2	55.2	....
C	♀	6	50.0-52.9	51.6	....
D	♂	17	52.9-58.2	55.4	2.64
D	♀	14	48.5-56.6	52.1	4.16
E	♂	10	50.9-56.9	55.1	3.40
E	♀	7	49.6-54.5	51.6	....
F	♀	2	51.9-52.6	52.3	....

A.—225 specimens: Dorsum mottled brown and cream, usually with a dark spot between the eyes and one or two dark V-shaped marks with the apex anteriorly on the back; 55 of these have a narrow cream-colored line from the tip of the snout to the vent and thence onto the posterior surfaces of the thighs. All are pustulate above; in most specimens the pustules form no pattern, but in some they tend to form a V in the scapular region.

B.—41 specimens: Dorsum pale tan or cream-color with brown mottling on flanks; a brown interorbital bar and a brown chevron in scapular region. Dorsum irregularly pustulate; in some specimens the pustules tend to form a V in the scapular region.

C.—12 specimens: Dorsum colored like "A", but having a broad yellow stripe narrowly bordered by black from the tip of the snout to the vent; in some specimens there is a narrow yellow stripe on the posterior surfaces of the thighs. The dorsum is irregularly pustulate.

D.—31 specimens: Dorsum variably streaked with cream-color or pale tan and brown; usually a broad cream-colored stripe from eyelid to groin bordered laterally by a somewhat narrower brown stripe; middorsal area cream-color and separated from dorsolateral cream-colored stripe by a brown stripe, or middorsal area brown with a cream-colored or yellow, narrow stripe from tip of snout to vent; a dark stripe from tympanum to flank; dorsal surfaces of heels creamy white to pale orange; anal patch brown. A dermal ridge from posterior edge of eyelid to rump; another ridge extends posteromedially from the eyelid; scattered pustules on the dorsum in some specimens.

E.—17 specimens: A narrow dark stripe from snout, through nostril and eye, over tympanum, to vent, enclosing a unicolor dorsum (reddish tan to yellowish tan in life); heels pale tan or yellow above; anal patch black. A faint dermal ridge from posterior edge of eyelid to rump, or part way to rump.

F.—2 specimens: Mottled brown and cream-color above; upper lips and upper arms white. A dermal fold from posterior edge of eyelid to rump; scattered pustules on dorsum.

Some of these color variants are assignable to names proposed by Taylor: "A" and "B" undoubtedly are *M. pygmaeus* (Taylor, 1936a); "C" probably is *M. pygmaeus*; "D" is referable to *M. minimus* (Taylor, 1940d) in most characteristics, although the coloration is more nearly like that of *M. lineatissimus* (Taylor, 1941a), a larger species characterized by a relatively long hind limb; "E" apparently is *M. imitator* (Taylor, 1942b); "F" is *M. albolabris* (Taylor, 1940d). Examination of series of these frogs from other parts of México shows a similar composition of color variants. Of 78 specimens from the Río Sarabia and the village of Sarabia in Oaxaca (UMMZ 115428-37), 57 are "A," six are "D," three are "E," and 12 are "F"; of 22 specimens from Teapa, Tabasco (UMMZ 113829), 11 are "A," five are "D," two are "E," and four are "F"; of 33 specimens from Potrero Viejo, Veracruz (USNM 115447-58, 115461-71, 116840-2, 116864-70), ten are "A," 13 are "E," and ten are "F"; of 31 specimens from La Esperanza, Chiapas (USNM 115477-9, 116827-39, 116849-63), 28 are "A" and four are "F."

It is highly doubtful if these color variants are actually distinct species. Goin (1950 and 1954) in his studies of inheritance of color pattern in West Indian species of the genus *Eleutherodactylus* has shown that similar color pattern variants come from the same clutch of eggs; furthermore, Goin has worked out the genetic ratios of certain of these variants. Heathwole (*in litt.*) obtained "normal" specimens and individuals having a broad middorsal stripe ("C" in figure 9) from a clutch of eggs of *Eleutherodactylus gollmeri*. The presence of a broad middorsal yellow stripe is common in *Eleutherodactylus rugulosus*.

Perhaps the most interesting aspect of variability in color pattern in Mexican eleutherodactylids is the parallelism between members of the *Eleutherodactylus rhodopis*-group and some members of *Microbatrachylus*. In the former group there are white-lipped individuals (*Eleutherodactylus beatae* Boulenger), individuals having a unicolor reddish or yellowish dorsum (*E. dorsoconcolor* Taylor), and individuals having a dorsal pattern of irregular longitudinal brown and cream-colored streaks (*E. venustus* Günther). In the humid forests of southern Veracruz, northern Oaxaca, and Chiapas members of both groups occur sympatrically. A proper understanding of the evolutionary significance of these variants in the two groups, as well as proper allocation of the presently recognized species, must await experimental evidence based on studies of the inheritance of color pattern. Nevertheless, at present it is apparent that certain characters, especially the nature of the dermal folds and pustules, and the color pattern, are of little taxonomic value in distinguishing "species" of *Microbatrachylus*. The data derived from a study of the large series from Arteaga, together with that from the other series examined, suggests that *Microbatrachylus albolabris, imitator, minimus,* and *pygmaeus* are morphotypes of one species. Of these names, *pygmaeus* is the oldest. Consequently *Microbatrachylus pygmaeus* has been used here for the series from Arteaga.

Although *Microbatrachylus hobartsmithi*, a species distinguished from all of the above by the presence of tubercles on the outer edge of the tarsus, is known from Michoacán northward into Nayarit, *Microbatrachylus pygmaeus* previously has not been known north of Guerrero, where it occurs in habitats similar to that in which it was collected at Arteaga.

ELEUTHERODACTYLUS AUGUSTI CACTORUM TAYLOR

Eleutherodactylus cactorum Taylor, Univ. Kansas Sci. Bull., 25:391, July 10, 1939.—20 miles northwest of Tehuacán, Puebla, México.

Eleutherodactylus augusti cactorum, Zweifel, Amer. Mus. Novitates, 1813:20, December 23, 1956.

Cherán; Coalcomán; Uruapan.

The few specimens indicate that this species occurs at moderate to high elevations in the state. The specimens from Cherán and Uruapan were obtained in pine forests; the specimen from Coalcomán was found on a rocky hillside covered with dense forest and located about 100 meters below the lower limits of the pine forest in the area. A specimen from Rancho Reparto (elevation 1850 meters) on the west slope of Cerro Barolosa was lost.

The specimen from Coalcomán (UMMZ 104728) is a juvenile having a snout-vent length of 25.0 mm. In life it was tan above, mottled with olive-green. The ventral surfaces were gray; the hind limbs were distinctly barred with yellow and brown, and the lips were barred with yellow and black.

ELEUTHERODACTYLUS OCCIDENTALIS TAYLOR

Eleutherodactylus occidentalis Taylor, Proc. Biol. Soc. Washington, 54:91, July 31, 1941.—Hacienda El Florencio, Zacatecas, México.

Arteaga (2); Cascada Tzararacua; Coalcomán (2); 19 km. SW
of Coire (3); La Placita (7); Los Reyes; Ostula (4); Pómaro (2).

The locality records for this species suggest that it is a member of a group of
animals, the distribution of which includes the western part of the Mexican Plateau and the Pacific lowlands. In Michoacán this frog has been collected in pine-
oak forest at Cascada Tzararacua and at Los Reyes, in arid scrub forest at Arteaga
and Coalcomán, and in tropical semi-deciduous forest on the lower Pacific slopes
of the Sierra de Coalcomán. On July 5, 1950, James Peters (1954:6) found calling
males at La Placita.

Most of the specimens are immature; four adult males have snout-vent lengths
of 30.9-33.0 (32.2) mm. In all specimens the first finger is noticeably longer than the
second; the inner metatarsal tubercle is large, flat, and cream-colored, contrasting
with the dark brown sole of the foot. When the hind limbs are adpressed, the heels
broadly overlap. Characteristically, a dark line extends from the snout, through
the eye, above the tympanum, to a point above the insertion of the forelimb. Usually there is a dark bar behind the tympanum, two dark brown bars from the eye
to the mouth and thence onto the lower jaw, and another dark bar on the upper
lip between the eye and nostril. One adult from Arteaga, an adult and a juvenile
from La Placita, and one juvenile each from Coire, Ostula, and Pómaro, have the
lower lip barred with dark brown and white, and have a white stripe extending
the length of the upper lip. In life the dorsum varies from dark gray or olive-brown
to tan or reddish brown.

This species belongs to a group containing two other populations that are currently recognized as species—*calcitrans*, known only from Omiltemi, Guerrero,
and *mexicanus*, reported from the mountains of Oaxaca. Another apparently un-
described member of this group has been collected in the mountains of northern
Puebla. The locality records indicate that the group inhabits the mountains on the
periphery of the Mexican Plateau, except in western México, where *Eleutherodactylus occidentalis* extends to the Pacific lowlands.

ELEUTHERODACTYLUS RUGULOSUS VOCALIS TAYLOR

Eleutherodactylus vocalis Taylor, Univ. Kansas Sci. Bull., 26:401,
November 27, 1940.—Hacienda El Sabino, Michoacán, México.

Arteaga (10); El Sabino (8); Salitre de Estopilas (3); Tumbiscatio (2); Tzitzio (2).

The distributional data on this frog in Michoacán indicate that it inhabits riparian situations in arroyos and canyons in the lower slopes of the Cordillera Volcánica and the Sierra de Coalcomán, where it has been taken at elevations only
below 1100 meters.

The dorsal color of living individuals from Arteaga varied from dark gray and
olive brown to tan and reddish brown. The iris was grayish brown. In contrast, individuals from Agua del Obispo, Guerrero, had pale golden eyes; specimens from
Matías Romero, Oaxaca, had gold eyes heavily flecked with gray; and individuals
from Volcán San Martin, Veracruz, had bronze eyes.

The use of the trinomial here is arbitrary. Frogs of the *Eleutherodactylus rugulosus* group in México (*rugulosus, avocalis,* and *vocalis*) exhibit only slight differences in size, proportions, and coloration (Duellman, 1958c:6). Furthermore, the named populations are allopatric. *Eleutherodactylus rugulosus vocalis*, as defined by Duellman (*loc. cit.*), occurs in the foothills of the Sierra Madre Occidental and associated ranges from central Sinaloa southward into Michoacán.

TOMODACTYLUS ANGUSTIDIGITORUM TAYLOR

Tomadactylus angustidigitorum Taylor, Univ. Kansas Sci. Bull., 26:494, November 27, 1940.—Quiroga, Michoacán, México.

Angahuan (6); Apo; Carapan (21); 19 km. S of Carapan (13); Cerro Tancítaro (12); Cherán; Corupu (14); Cuseño Station (14); Opopeo (3); Paracho (11); Parícutin (2); Pátzcuaro (3); Quiroga (59); San Juan de Parangaricutiro (16); Tancítaro (25); Uruapan (8); Zacapu (11).

This species is indigenous to the pine-oak forests on the southern rim of the Mexican Plateau, and has been collected at elevations from 1500 to 2500 meters. Males have been observed to call from rocks, rock fences, clumps of grass, and low bushes; the call is a single "peep." At San Juan de Parangaricutiro numerous specimens were found in the daytime beneath adobe bricks and lava on the volcanic ash derived from Volcán Parícutin; at Paracho individuals were found by day beneath rocks in a pine forest.

In most specimens the dorsum is dark reddish brown, and the prominent inguinal glands are cream-color or pale orange (Pl. 3, Fig. 1). Of eight individuals collected at Paracho, one was reddish brown, two were pinkish tan, three were dark brown, and two were black.

TOMODACTYLUS FUSCUS DAVIS AND DIXON

Tomodactylus fuscus Davis and Dixon, Herpetologica, 11:157, July 15, 1955.—1.5 miles southeast of Huitzilac, Morelos, México.

Los Cantiles (2); 28 km. E of Morelia.

The range of this species includes the Sierra Ajusco in México and Morelos and thence westward to the Serranía Ucareo in Michoacán. The specimen from 28 kilometers east of Morelia was found in an oak forest on a steep hillside at an elevation of 2100 meters. One from Los Cantiles was calling from a steep cliff at an elevation of 2200 meters in pine-oak forest. This specimen (UMMZ 119156) in life had a pale olive-brown dorsum with irregular dark brown mottling and transverse bars on the limbs. The interorbital bar, the upper arms, and the tips of the dorsal pustules were pale orange; the iris was pale grayish gold (Pl. 3, Fig. 2).

TOMODACTYLUS NITIDUS NITIDUS (PETERS)

Liuperus nitidus Peters, Monats. Akad. Wiss. Berlin, p. 878, 1869.—Izúcar de Matamoras, Puebla, México.

Tomodactylus amulae Günther, Biologia Centrali-Americana, Reptilia and Batrachia, p. 219, April, 1900.—Amula, Guerrero, México.

Tomodactylus nitidus nitidus, Dixon, Texas Jour. Sci., 9:385, December, 1957.

Copuyo (15); Tuxpan (8); Tzitzio (11).

One specimen from Tzitzio (UMMZ 99155) was referred to *Tomodactylus nitidus petersi* by Dixon (1957:390). A re-examination of this specimen, and examination of ten others from the same locality (UMMZ 121571) reveals that the relatively small size of the tympanum and absence of dense ventral spotting place these specimens closer to *T. nitidus nitidus* than to *T. nitidus petersi*.

The specimens from Tuxpan (UMMZ 114303-4) had in life a gray to olive tan ground color with dark olive-green markings, bright yellow thighs with olive-green transverse bands, yellowish tan shanks with olive-green bars, yellow groin, white inguinal glands with black markings, grayish white belly with scattered brownish black spots in some specimens, and a deep golden iris (Pl. 4, Fig. 1). These specimens were found calling from bushes in a rocky field at an elevation of 1800 meters. The call is a high-pitched "pee-ee-eep."

TOMODACTYLUS NITIDUS ORARIUS DIXON

Tomodactylus nitidus orarius Dixon, Texas Jour. Sci., 9:392, December, 1957.—4.5 miles southwest of Tecolapa, Colima, México.

La Placita (3); Pómaro.

These specimens, referred to *Tomodactylus petersi* by Duellman (1954b:5), were included in *T. nitidus orarius* by Dixon (1957:392). Color notes based on living individuals from Tecolapa, Colima (UMMZ 114312 and 116922), are: gray above mottled with brown; venter dirty white; anterior and posterior surfaces of thighs bright yellow; iris pale golden (Pl. 4, Fig. 2). The call is a soft "braa" usually followed by three high notes: "braaa-eep-ee-eep." In Michoacán this subspecies has been found only in the coastal region and the lower foothills of the Sierra de Coalcomán, an area in which it replaces *Tomodactylus nitidus petersi*. This is the only *Tomodactylus* known to inhabit coastal lowlands.

TOMODACTYLUS NITIDUS PETERSI DUELLMAN

Tomodactylus petersi Duellman, Occ. Pap. Mus. Zool. Univ. Michigan, 560:5, October 22, 1954.—Coalcomán, Michoacán, México.

Tomodactylus nitidus petersi, Dixon, Texas Jour. Sci., 9:390, December, 1957.

Aguililla; Apatzingán (8); Cascada Tzararacua: Charapendo (5); Coalcomán (5); 18 km. E of Dos Aguas (6); El Sabino (5); La Playa (2); Jiquilpan; Uruapan (2); Volcán Jorullo; Zamora.

In life, specimens from Apatzingán (UMMZ 114308-9) varied in dorsal color from grayish tan to pale brown; the dorsal markings were olive green. The thighs and groin were yellowish orange; the iris was pale golden, and the vocal sac was purplish gray (Pl. 5, Fig. 1). Measurements for 13 adult males from the Tepalcatepec Valley are: snout-vent length, 21.9-26.8 (24.3); tibia length, 8.4-9.9 (9.3); head width, 7.2-9.2 (7.8); head length, 7.6-8.7 (8.2).

At Apatzingán and Charapendo in the Tepalcatepec Valley males were found calling from rocks and bushes in open arid tropical scrub forest. The call, a triple note "peep-ee-eep," is repeated once every 90 to 135 seconds. *Tomodactylus nitidus petersi* probably ranges throughout the Tepalcatepec Valley and surrounding foothills. Dixon (1957:392) referred the specimens from Zamora, Jiquilpan, and Uruapan to this subspecies. Uruapan is near the lower limits of the pine forest on the slopes of the Cordillera Volcánica; Zamora and Jiquilpan are on a low part of the Mexican Plateau southeast of Lago de Chapala.

TOMODACTYLUS RUFESCENS DUELLMAN AND DIXON

Tomodactylus rufescens Duellman and Dixon, Texas Jour. Sci., 11:78, March, 1959.—Dos Aguas, Michoacán, México.

Dos Aguas (14); 18 km. E of Dos Aguas (6).

Fourteen specimens from the pine-oak forests around Dos Aguas (UMMZ 118503-10, 121498-9) have reddish brown dorsal color and a narrow cream-colored middorsal line (Pl. 5, Fig. 2). Twelve of these specimens are adult males having snout-vent lengths of 20.7 to 24.6 (22.5) mm. One female has a snout-vent length of 24.8 mm., and one juvenile has a snout-vent length of 14.5 mm. Six specimens are from a region of mixture of pine-oak forest and arid tropical scrub forest at 18 kilometers east of Dos Aguas (UMMZ 121497, 121500). All are males having snout-vent lengths of 18.0 to 22.6 (20.7) mm. The dorsum is tan marked with black; the thighs are yellowish orange.

The specimens from 18 kilometers east of Dos Aguas were found on July 22, 1960, by Floyd L. Downs and John Winklemann, who collected calling males of *Tomodactylus rufescens* and *Tomodactylus nitidus petersi* at the same locality. Downs (*personal communication*) stated the call was a single note. At Dos Aguas I heard *T. rufescens* give two calls, one a single "peep," the other a triple note—"pee-ee-eep."

In the higher parts of the Sierra de Coalcomán *Tomodactylus rufescens* seems to fill the same niche as *T. angustidigitorum* does in the Cordillera Volcánica. At lower elevations in their respective mountain ranges the species occur sympatrically with *T. nitidus petersi.*

DIAGLENA RETICULATA TAYLOR

Diaglena reticulata Taylor, Univ. Kansas Sci. Bull., 28:60, May 15, 1942.—Cerro Arenal, Oaxaca, México.

Nueva Italia (3); Ostula (7).

Until recently frogs of the genus *Diaglena* were known only from a few specimens from southern Sinaloa (*Diaglena spatulata*) and from the Pacific lowlands of the Isthmus of Tehuantepec (*Diaglena reticulata*). Peters (1955a) reported specimens from Ostula, Michoacán, and compared these specimens with one *D. reticulata* from Tehuantepec, Oaxaca, and four *D. spatulata* from Sinaloa. This comparison showed that the specimens from Michoacán, although showing some minor differences from *D. reticulata*, are closer to that species than to *D. spatulata*. Subsequent to Peters' work, series of both species of *Diaglena*, including additional specimens from Michoacán and from Colima, have been collected, and a more qualified comparison is now possible.

In comparing specimens of *D. spatulata* from southern Sinaloa (UMMZ 115322) with specimens of *D. reticulata* from Tehuantepec, Oaxaca (UMMZ 115321), the differences noted by Taylor (1942c:60) were found to be constant. But specimens from Ostula, Michoacán (UMMZ 104418), and five individuals from Colima (TNHC 26379-83) were found to be intermediate in certain characters. The skin of the dorsum in *D. reticulata* is granular; that in *D. spatulata* is smooth. The skin in specimens from Ostula and Colima is slightly granular. The dorsal ground color of *D. reticulata* is yellowish brown with dark reticulations; the dorsal ground color of *D. spatulata* is olive-green. Specimens from Ostula and Colima most closely resemble those from Tehuantepec in coloration, but the reticulations are more coarse, and the ground color has an olive-green tint. *Diaglena reticulata* also differs from *D. spatulata* in having a larger over-all size, slightly broader head, a narrower interorbital distance, and a more pointed snout with a deeper labial shelf (Table 4). The specimens from Ostula and Colima are intermediate between *D. reticulata* from Oaxaca and *D. spatulata* from Sinaloa in body proportions.

Of three specimens from the Tepalcatepec Valley (JRD 5991-3), only two are suitable for measuring. These specimens are smaller than adults from the coastal areas and have broader heads and snouts, but narrower interorbital distances, than specimens in the other samples (Table 4). The texture of the skin is like that of specimens from Ostula and Colima. The coloration resembles that of *D. reticulata*, but the reticulations are bold and form indistinct bands on the hind limbs.

TABLE 4.—COMPARISON OF FOUR CHARACTERS IN FIVE SAMPLES OF DIAGLENA. (ALL DATA ARE FOR MALES; MEANS GIVEN IN PARENTHESES BELOW RANGES.)

Locality	Number of specimens	Snout-vent length	Head width	Interorbital distance	Internarial distance
			Snout-vent length	Head width	Head width
Oaxaca	9	71.1-87.5	25.4-29.1	63.0-71.4	11.9-13.8
		(80.7)	(27.9)	(67.1)	(12.9)
Coast of Michoacán	5	72.0-79.2	24.3-27.2	67.0-73.8	13.7-14.4

		(74.8)	(25.6)	(71.4)	(14.1)
Colima	4	71.7-79.6	26.1-28.6	70.5-75.3	16.0-17.9
		(74.8)	(27.4)	(72.0)	(16.6)
Tepalcatepec Valley	2	63.0-65.4	28.3-32.2	57.3-62.4	17.0-20.2
		(64.2)	(30.3)	(59.9)	(18.6)
Sinaloa	11	71.9-81.3	24.0-27.3	70.5-78.1	15.0-17.3
		(77.3)	(25.7)	(73.4)	(16.1)

All specimens from Michoacán and Colima more closely approach *Diaglena reticulata* than *D. spatulata*. The acquisition of additional specimens, especially from the area between Sinaloa and Colima and from Guerrero, is necessary to determine the relationships among the various populations known at present. Both species of *Diaglena* inhabit tropical scrub forest; none has been found in the more humid and tropical semi-deciduous forests. Humid forest replaces the scrub forest in the lowlands of southern Nayarit and northern Jalisco; possibly this forest acts as a barrier to the distribution of *Diaglena* and thus serves as a divider between the ranges of *D. spatulata* to the north and *D. reticulata* to the south.

PTERNOHYLA FODIENS BOULENGER

Pternohyla fodiens Boulenger, Ann. Mag. Nat. Hist., ser. 5, 10:326, 1882.—Presidio, Sinaloa, México.

Nueva Italia (2).

These specimens (JRD 5994-5) were found on the road near Nueva Italia during a heavy rain on the night of August 25, 1960, by James R. Dixon. Both are females having snout-vent lengths of 64.0 and 59.0 mm. They are typical of the species as it is known from Sinaloa, Nayarit, Jalisco, and Colima.

These specimens constitute the southernmost record for the species, which ranges in semi-arid habitats from southern Arizona southward along the Pacific lowlands of México to Colima and inland on the Mexican Plateau in Jalisco.

PHYLLOMEDUSA DACNICOLOR COPE

Phyllomedusa dacnicolor Cope, Proc. Acad. Nat. Sci. Philadelphia, 16:181, September 30, 1864.—Colima, Colima, México. Funkhouser, Occ. Pap. Nat. Hist. Mus. Stanford Univ., 5:37, April 1, 1957.

Agalychnis alcorni Taylor, Copeia, no. 2:31, June 2, 1952.—Río Tepalcatepec, 17 miles south of Apatzingán, Michoacán, México.

Agalychnis dacnicolor, Duellman, Herpetologica, 13:29, March 30, 1957.

Phyllomedusa alcorni, Funkhouser, Occ. Pap. Nat. Hist. Mus. Stanford Univ., 5:30, April 1, 1957.

Aguililla (13); Apatzingán (7); Charapendo; Coahuayana (3); Coalcomán (54); El Sabino; Huetamo Road (2); La Orilla; La Placita; Nueva Italia (4); 32 km. E of Neuva Italia (2); Río Cancita, 14 km. E of Apatzingán; Río Tepalcatepec, 27 km. S of Apatzingán; Salitre de Estopilas (2); Tzitzio (4).

This large tree frog has been found only in the lowlands below elevation of about 1000 meters, usually in arid tropical scrub forest. Calling males were heard on rainy nights throughout the rainy season; in nearly every instance both males and females were found in low trees and bushes. On summer nights when there had been no rain, adults were found sitting on bushes in the scrub forest.

At Coalcomán on July 1, 1955, a chorus was heard at midday. About forty *Phyllomedusa dacnicolor* were found in one guayava bush at the edge of a recently dried pond. Individual males were calling; clasping males were silent. The call is a barking groan. Fifteen individual egg masses were hanging from branches and leaves in tear-drop fashion. Each egg mass contained 100 to 350 pale green eggs, located only in the exterior part of the clear gelatinous mass. Two composite egg masses appeared to have been made up by egg deposition on the part of three to five females (Pl. 2, Fig. 2).

As shown by Duellman (1957a), the characters used by Taylor (1952) to diagnose *Phyllomedusa alcorni* are sexually dimorphic. Funkhouser (1957) apparently was unaware of this sexual dimorphism, for she recognized *P. alcorni* and *P. dacnicolor* as distinct species.

PHRYNOHYAS INFLATA (TAYLOR)

Acrodytes inflata Taylor, Univ. Kansas Sci. Bull., 30:64, June 12, 1944.—La Venta, Guerrero, México.

Phrynohyas inflata, Duellman, Misc. Publ. Mus. Zool. Univ. Michigan, 96:19, February 1, 1956.

Phrynohyas corasterias Shannon and Humphrey, Herpetologica, 13:15, March 30, 1957.—4.8 miles east of San Blas, Nayarit, México.

Barranca de Bejuco.

One specimen of this large species was collected in 1951; it was found on a low branch in tropical semi-deciduous forest at an elevation of 65 meters. In life there were olive-gray blotches on a pale gray dorsum; the iris was a dark golden color.

This species, which is known from only a few specimens, seems to be restricted to the coastal lowlands and low foothills from Guerrero northward to Nayarit. Shannon and Humphrey (1957) described *Phrynohyas corasterias* from Nayarit. Their description was based on a small female having a snout-vent length of 34.4 mm. The new species was diagnosed as differing from *P. inflata* in having less webbing on the feet, a poorly developed supratympanic fold, a more pustulate dorsum, and marked differences in dorsal pattern, color, and nature of antebrachial banding. The significance of the webbing was questioned by Shannon and

Humphrey. The nature of the supratympanic fold and dorsal pustules changes with age (Duellman, 1956a:31). *Phrynohyas inflata* is known to attain a snout-vent length of 95 mm. Dermal structures that undergo ontogenetic change are of little importance in comparing a juvenile with a large adult. The only significant difference in color pattern between *P. inflata* and *P. corasterias* is the presence of wide transverse bands on the limbs of the latter. In this respect *P. corasterias* approaches *P. latifasciata*, a species known only from two specimens from southern Sinaloa. The acquisition of additional specimens from Jalisco, Nayarit, and Sinaloa may show that *P. inflata* and *P. latifasciata* are conspecific, as suggested by Duellman (1956a:21). Nonetheless, the specimen on which the description of *P. corasterias* was based is not sufficiently different from the known specimens of *P. inflata* to warrant specific recognition.

HYLA ARENICOLOR

Hyla arenicolor Cope, Jour. Acad. Nat. Sci. Philadelphia, ser. 2, 6:84, July, 1866.—Northern Sonora, México. Type locality restricted to Santa Rita Mountains, Pima County, Arizona, by Smith and Taylor (1950a:354).

Agua Cerca; Cascada Tzararacua (3); Chinapa; Cojumatlán; Dos Aguas; El Sabino (25); El Espinal; Lago de Camécuaro; Lombardia (2); Tupátaro; Zinapécuaro.

Altitudinally this frog ranges from 500 to 2100 meters; although the environments in which it has been found vary from open arid tropical scrub forest to pine forest, it usually is found near rocky streams in these habitats. There is great disparity in size between specimens from the mountains and those from the Tepalcatepec Valley. Seven males from elevations in excess of 1400 meters have an average snout-vent length of 34.7 mm.; nine from elevations below 1000 meters have an average snout-vent length of 49.1 mm. In life a male collected at night at Lombardia (UMMZ 112846) had dark brown spots on a grayish brown dorsum; the groin, anterior and posterior surfaces of the thighs, and ventral surfaces of the hind limbs and palms were yellowish orange. The belly and tips of digits were white; the vocal sac was purplish brown, and the iris was dark grayish gold. In contrast, a specimen obtained in the daytime at Chinapa (UMMZ 119204) had indistinct gray spots on a pale ashy gray dorsum; the flash colors were yellow. After dark the spots were dark olive-brown on a grayish brown dorsum.

Two males were found calling from a rocky stream near Lombardia on July 12, 1955. The call is a nasal "ah-ah-ah-ah."

HYLA BAUDINI DUMÉRIL AND BIBRON

Hyla baudinii Duméril and Bibron, Erpétologie générale, vol. 8:564, 1841.—México. Type locality restricted to Córdoba, Veracruz, México, by Smith and Taylor (1950a:346).

Aguililla (5); Apatzingán (30); Arteaga; Buena Vista; Charapendo; Coahuayana; Cofradía (4); El Sabino (12); La Placita; La

Playa; Maruata; Nueva Italia (3); 32 km. E of Nueva Italia (2); Ostula (4); Río Tepalcatepec, 25 km. S of Apatzingán (3); Salitre de Estopilas; San José de la Montaña (2); Tumbiscatio; Tzitzio.

This tree frog is widespread in the coastal lowlands and in the Tepalcatepec Valley up to elevations of about 1200 meters. It is found in numbers in the early part of the rainy season, at which time males were heard calling from bushes and trees along ditches and temporary ponds. The call is a loud nasal "waank-waank-waank." One individual that was emitting a long and unusually high-pitched call was found to have one hind limb engulfed by a *Leptodeira maculata*.

When active at night these frogs usually are pale tan to reddish brown above with dark brown markings. A specimen found sitting on a maguey plant in the daytime was pale ashy gray with a pale green upper lip.

HYLA BISTINCTA COPE

Hyla bistincta Cope, Proc. Amer. Philos. Soc., 17:87, 1877.—Veracruz, México. Type locality restricted to Acultzingo, Veracruz, México, by Smith and Taylor (1950a:346).

Cerro San Andrés; Dos Aguas (2); Los Conejos (3); Uruapan (50).

In the Parque Nacional at Uruapan this species was found in abundance during the day. The frogs hide in an entanglement of vines and vegetation overhanging several small spring-fed streams. Tadpoles were in the rocky streams, and metamorphosing young were on vegetation at the edges of the streams.

In life the dorsum is greenish tan with brown mottling; in some individuals the entire dorsum is dark chocolate brown. The flanks are pale lemon yellow barred with lavender-brown. Notes on the color of a living frog from Dos Aguas (UMMZ 119193) are: Dorsal ground color a medium shade of brown with dark brown flecks; flanks black with silvery white and pale yellow spots; belly pale yellowish white; throat mottled with grayish brown; iris pale copper color.

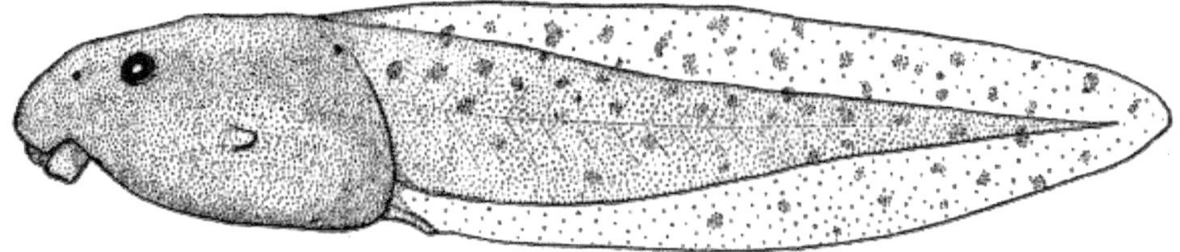

**Fig. 8. Tadpole of Hyla bistincta (UMMZ 115231) from Uruapan, Michoacán. ×
2.**

Description of Tadpole: Body somewhat depressed; maximum width of body slightly more than one-half of body length. Nostrils placed dorsolaterally and directed anteriorly, situated about midway between tip of snout and eye. Eyes of moderate size, dorsolateral in position and directed upwards. Tail about twice as long as body, thrice as long as deep, and tapering gradually to a rounded tip. Tail-musculature not extending to tip of tail fin. Spiracle sinistral, lateral, and situated at midbody. Vent dextral; the cloacal tube extending along ventral part of tail for a distance equal to about one-eighth of body length (Fig. 8). Average body length of six tadpoles with small hind limb buds, 19.5 mm.; tail length, 38.3 mm. Mouth ventral, its width equal to about two-thirds of greatest width of body. Lips bordered by two rows of small papillae; row of larger papillae between upper lip and outer upper tooth-row, similar row between lower lip and outer lower tooth-row; laterally these rows degenerating into numerous small papillae. Horny beaks well developed; upper beak moderately arched and deeply indented; lower beak slightly indented. Serrations of beaks blunt and peglike, moderately developed on both beaks, but slightly stronger on lower one. Tooth-rows 2/3; upper rows nearly equal in length and slightly longer than lower rows, which are subequal in length; inner upper tooth row interrupted medially by rounded notch; inner lower tooth-row turned downward laterally; teeth in all rows about equal in size, but decreasing in length laterally (Fig. 9).

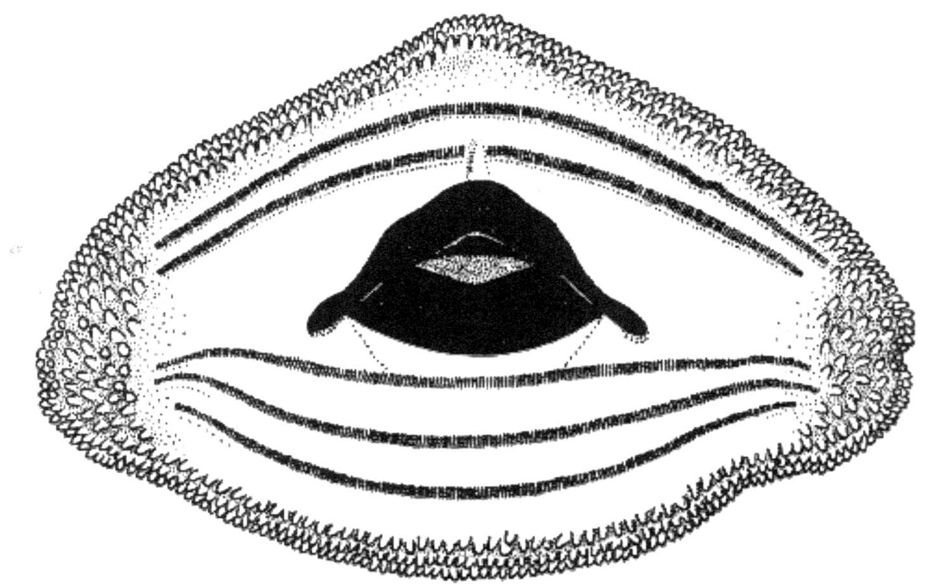

Fig. 9. Mouthparts of larval Hyla bistincta (UMMZ 115231) from Uruapan, Michoacán. × 15.

Color in formalin: pale grayish brown dorsally and laterally; pale gray ventrally; tail-musculature brown; tail-fin translucent with scattered melanophores most numerous on upper fin.

In most details these tadpoles resemble those of *Hyla robertsorum* described by Rabb and Mosimann (1955).

Four metamorphosing young have snout-vent lengths of 23.0-23.5 (23.2); a completely metamorphosed individual has a snout-vent length of 24.8 mm.

In Michoacán this stream-breeding hylid occurs at elevations of 1,600 to 2,400 meters in the Sierra de Coalcomán and in the mountains rising from the Mexican Plateau.

HYLA EXIMIA BAIRD

Hyla eximia Baird, Proc. Acad. Nat. Sci. Philadelphia, 7:61, October 20, 1854.—Valley of México. Type locality restricted to Coyoacán, Distrito Federal, México, by Smith and Taylor (1950a:329).

Hyla microeximia Maslin, Herpetologica, 13:81, July 10, 1957.— 3 miles northwest of Jocotepec, Jalisco, México.

Ciudad Hidalgo (36); Cuitzeo; 29 km. NW of Jacona; Jiquilpan (2); Lago de Camécuaro (2); Lago de Pátzcuaro (129); Los Reyes; Morelia; Sahuayo (3); San Gregorio (63); Tangamandapio (4); Temazcal (26); Tupátaro; Tuxpan (15); Undameo (2); Uruapan (20); Zacapu; Zamora (27); Zinapécuaro (10).

More than 80 per cent of the specimens from Michoacán have brown spots between the lateral and dorsolateral dark stripes, and more than 50 per cent have spots between the dorsolateral stripes, at least posteriorly. In comparison with specimens from the Valley of México, those from Michoacán have more distinct dorsolateral stripes that extend farther anteriorly, sometimes to the eyelid, and in this respect are more nearly like those from Jalisco and Nayarit (Taylor, 1939b:425). Some specimens from the western part of Michoacán possess certain characters used by Maslin (1957:81) to distinguish *Hyla microeximia* from *H. eximia*; nevertheless, the variation is such that two species cannot be distinguished in Michoacán. Four series of freshly preserved specimens have been studied in detail; in the discussion below they are arranged from west to east; the measurement is for snout-vent length of ten males from each sample:

Zamora.—Twenty-two specimens (UMMZ 102083), 24.0-27.6 (26.1) mm. Dorsolateral dark stripe, or row of dashes, present in all specimens; dark spots in lateral and dorsal green fields; lateral dark stripe confluent with dorsolateral stripe posteriorly in 18 specimens; white line not extending to groin.

Temazcal.—Thirty-five specimens (UMMZ 119162), 26.5-31.1 (28.2) mm. Dorsolateral dark stripe of row of spots present only posteriorly in most; both dorsolateral stripes and dorsal spots lacking in four specimens; heavy spotting dorsally in three others; lateral and dorsolateral dark stripes confluent posteriorly in three; lateral white stripe extending to groin in 16 specimens.

Ciudad Hidalgo.—Thirty-six specimens (UMMZ 119163), 26.4-30.9 (28.2) mm. Dorsolateral dark stripe or row of spots present only posteriorly in most; no brown spots in the green fields of many specimens; large brown inguinal spot in most specimens; heavy spotting dorsally in four; lateral and dorsolateral dark stripes confluent posteriorly in five; lateral white line extending to groin in most specimens.

Tuxpan.—Fifteen specimens (UMMZ 115227), 28.7-33.0 (30.5) mm. Dorsolateral dark stripe or row of dashes in all specimens; dark spots in lateral green fields, at least posteriorly in most; dark spots posteriorly in the dorsal green field in five; lateral dark stripe separated from dorsolateral stripe in all specimens; lateral white line extends to the groin in all specimens.

As can be seen from the above descriptions, the distinguishing characters of *Hyla microeximia*—confluence of lateral and dorsolateral dark stripes posteriorly, extent of lateral white stripe, and distribution of dark spots dorsally—are found in individuals from all of the populations sampled. In the samples from western Michoacán there is a higher incidence of *microeximia*-like frogs than in those from other parts of the state. *Hyla eximia* is a wide-ranging species varying greatly geographically and individually. A thorough review of the species and related members of the *Hyla eximia*-group is necessary before certain populations can justifiably be segregated as subspecies or species.

In Michoacán *Hyla eximia* has been collected in mesquite-grassland, pine-oak forest, and cultivated areas on the Mexican Plateau from 1500 to 2300 meters; apparently it is absent from the Sierra de Coalcomán. This is the most abundant frog on the southern part of the Mexican Plateau; in the rainy season breeding choruses are found in temporary pools and in the marshes adjacent to the permanent lakes.

HYLA LAFRENTZI MERTENS AND WOLTERSTORFF

Hyla lafrentzi Mertens and Wolterstorff, Zool. Anz., 84:235, August 25, 1929.—Desierto de los Leones, Distrito Federal, México.

Cerro San Andrés (26); Opopeo (9).

In March, 1949, James A. Peters collected this species at elevations of 2400 to 2800 meters on the west slope of Cerro San Andrés. The frogs were found beneath logs and rocks in a damp canyon in coniferous forest. Among the juveniles in this series is a completely transformed individual (UMMZ 102093) having a snout-vent length of 14.5 mm. Five adults have snout-vent lengths of 36.2-39.5 (38.0) mm. *Hyla lafrentzi* has noticeably longer hind limbs than *H. eximia*; in the former, when the hind limb is brought forward along the body, the tibiotarsal articulation extends to the snout. There are dark transverse bands on the hind limbs; the dorsolateral stripe is broken into an anterior and a posterior segment, and the latter is narrowly bordered by white in most specimens.

Hyla lafrentzi occurs at higher elevations than any other frog in Michoacán; the locality records from throughout the range indicate that it is restricted to pine and pine-fir forests. In these habitats it replaces *Hyla eximia*, which inhabits the lower pine-oak forests and mesquite-grassland on the Mexican Plateau. Ponds are absent at places where *Hyla lafrentzi* has been collected; possibly the eggs are laid in streams.

HYLA SMARAGDINA TAYLOR

Hyla smaragdina Taylor, Copeia, No. 1:18, March 30, 1940.—6 kilometers east of Cojumatlán, Michoacán, México.

Hylella azteca Taylor, Proc. Biol. Soc. Washington, 56:49, June 16, 1943.—Tepoztlán, Morelos, México.

Cojumatlán (30); Copuyo (7); 18 km. E of Dos Aguas (22); Ostula (8); Pómaro (3); Sahuayo; Salitre de Estopilas (7).

Taylor (1940a:18) diagnosed this species as having few or no vomerine teeth, no vocal sac, a rather broad and flat head, two large tubercles below the anus, a granular venter, and a green dorsum in life. The specimens on which the description was based are either immature or non-breeding individuals; all were collected from bromeliads growing on cacti near Cojumatlán. Another small, flat-headed hylid from Tepoztlán, Morelos, was described and diagnosed by Taylor (1943b:49) as differing from *Hyla smaragdina* in having a vocal sac and a broader head. This specimen was named *Hylella azteca*. Specimens from the coastal region of Michoacán and Colima were referred to *Hylella azteca* by Peters (1954:7) and Duellman (1958c:8).

Comparison of topotypic *Hyla smaragdina* and the holotype of *Hylella azteca* (UIMNH 25044) with the several series of specimens from Michoacán has resulted in the conclusion that all pertain to only one species. Although the type series of *Hyla smaragdina* consists of immature specimens, the males in that series do possess vocal sacs. Since these were not breeding individuals, the sacs are not well developed. The characters of the anal tubercles and the relative width of the head are of no value in separating the two species. The apparently aestivating individuals comprising the type series of *Hyla smaragdina*, and the type of *Hylella azteca*, which also was found in a bromeliad, were green in life. Of the calling males found on the coast of Michoacán, most were yellowish tan when found; two were pale green, but soon changed to pale tan. Calling males from Copuyo and Dos Aguas were pale yellowish tan. Therefore the color of the dorsum is of little significance in distinguishing the two named populations.

Males of *Hyla smaragdina* have been found calling in the months of June and July from rocky streams; the call is a nasal "haah-haah-haah," repeated quickly and constantly for as long as 30 seconds. As pointed out by Duellman (1958c:9), this breeding behavior is unlike that suggested by Taylor (1943b:51). In Michoacán *Hyla smaragdina* has been found in tropical semi-deciduous forest, oak forest, and mesquite-grassland at elevations from 150 to 1500 meters.

HYLA SMITHI BOULENGER

Hyla smithi Boulenger, Zool. Rec. Reptilia and Batrachia, 38:33, 1902.—Cuernavaca, Morelos, México.

Aguililla (14); Apatzingán (104); Arteaga; Charapendo (5);

Coalcomán (11); El Sabino (44); La Playa (6); Lombardia (2); Nueva Italia (8); Playa Azul; Salitre de Estopilas (2).

This small hylid is abundant in the Tepalcatepec Valley to elevations of about 1000 meters; it was found infrequently on the coastal lowlands. Males call from bushes in and around flooded fields and ditches, from grasses and small herbs in the water and from vegetation overhanging small streams. The call consists of a series of short, high notes, somewhat reminiscent of a katydid's song. In the dry season occasional males were heard calling from irrigated fields near Apatzingán. In the daytime individuals were found in the axils of leaves of the elephant-ear plants (*Xanthosoma*).

In living individuals the dorsal color usually is uniform pale yellow; often the lateral white stripe is barely visible. The vocal sac is bright yellow, and the iris is pale gold. In some individuals there are scattered dark brown spots or flecks on the back and upper surfaces of limbs. Twenty males from Apatzingán have the following measurements: snout-vent length, 22.8-26.0 (25.0) mm., tibia length, 10.7-13.6 (12.6) mm.; head width, 7.2-8.0 (7.6) mm., head length, 7.1-8.1 (7.7) mm.

HYPOPACHUS CAPRIMIMUS TAYLOR

Hypopachus caprimimus Taylor, Univ. Kansas Sci. Bull., 26:526, November 27, 1940.—Agua del Obispo, Guerrero, México.

Buena Vista; Copuyo (6); Charapendo (3); Cofradía; Jaramillo; Jungapeo; San Salvador; Tuxpan.

Specimens of *Hypopachus* from the Balsas drainage in Michoacán have characters consistent with topotypic *H. caprimimus*. Eleven specimens from the southern edge of the Mexican Plateau all have the flanks darker than the dorsum, a distinct and continuous dark stripe from the occiput to the groin, a large dark spot in the inguinal region, and a pair of dark transverse stripes on the thigh and shank (Pl. 6, Fig. 1). With the exception of three specimens from Charapendo, all have a predominantly brown venter with round, cream-colored spots.

Peters (1954:8) referred specimens from Buena Vista and San Salvador to *Hypopachus oxyrrhinus*. He stated that the specimen (BMNH 1914.1.28.150) from San Salvador had flanks much darker than the dorsum and a well-defined continuous stripe from the occiput to the groin; this specimen has the characters of *H. caprimimus*. The specimen (BMNH 1914.1.28.151) from Buena Vista resembles *H. oxyrrhinus* in some characters, but it is not like *H. oxyrrhinus ovis* on the Mexican Plateau in Michoacán. The specimen has paired transverse stripes on the hind limbs as does *H. caprimimus*, and is here referred to that species.

In Michoacán this species has been collected in arid tropical scrub forest at elevations of 200 to 1800 meters in the northern foothills of the Sierra de Coalcomán, the Tepalcatepec and Tuxpan valleys, and on the lower slopes of the Cordillera Volcánica. Calling males have been found along streams. One specimen from Charapendo was regurgitated by a *Leptodeira maculata*.

HYPOPACHUS OXYRRHINUS OVIS TAYLOR

Hypopachus ovis Taylor, Univ. Kansas Sci. Bull., 26:520, November 27, 1940.—Tepic, Nayarit, México.

Hypopachus oxyrrhinus ovis, Shannon and Humphrey, Herpetologica, 14:89, July 23, 1958.

Emiliano Zapata; 30 km. NW of Jacona (2); 10 km. NE of Pátzcuaro (2); Tangamandapio (16); 24 km. W of Zamora (16).

Thirty-seven specimens from the Mexican Plateau in northwestern Michoacán agree well with the diagnosis of *Hypopachus oxyrrhinus ovis* by Shannon and Humphrey (1958). With the exception of one specimen from Tangamandapio, all have dark bellies extensively mottled or spotted with cream-color. Most of the specimens have some form of an irregular, usually broken, dark line from the occiput to the groin. In eight specimens there is no line or linear arrangement of spots; instead the dorsum is spotted or flecked with dark brown. The ground color of the dorsum and flanks varies from dull reddish brown to grayish brown; cream-colored spots are evident on the flanks and posterior surfaces of the thighs in all specimens (Pl. 6, Fig. 2).

In comparison with 14 specimens from Quesería, Colima (UMMZ 80001-2), individuals from the Mexican Plateau have a darker venter with bolder markings, and a more mottled dorsum.

In Michoacán this species has been taken between 1500 and 2200 meters on the Mexican Plateau, where it inhabits mesquite-grassland and cultivated areas.

RANA DUNNI ZWEIFEL

Rana dunni Zweifel, Copeia, no. 2:78, July 15, 1957.—Lago de Pátzcuaro, Michoacán, México.

Lago de Pátzcuaro (23); Río de Morelia, near Undameo (8).

Aside from the type series of this species, there are in the Museum of Zoology at the University of Michigan six specimens taken from "tanks" at the limnological station at Pátzcuaro by Paul S. Martin in 1948, and eight specimens found in shaded ditches along the Río de Morelia by Robert R. Miller on April 4, 1957. The Río de Morelia flows into Lago de Cuitzeo; this drainage is separated from Lago de Pátzcuaro by a chain of hills about 2400 meters in elevation. Dr. Richard G. Zweifel has examined these specimens and has informed me that, although they differ slightly from typical *Rana dunni*, they are much closer to that species than to *Rana montezumae*.

RANA MEGAPODA TAYLOR

Rana megapoda Taylor, Univ. Kansas Sci. Bull., 28:310, November 12, 1942.—Chapala, Jalisco, México.

La Palma (8).

These specimens (USNM 113998-114005) are from the marshes along the southeastern shore of Lago de Chapala. Five females have snout-vent lengths of

124.0-138.1 (131.5), and one male has a snout-vent length of 110.2 mm. Two juveniles have snout-vent lengths of 49.7 and 56.3 mm. The coloration of the juveniles is more bold than that of the adults. The body proportions of these specimens agree with those presented by Zweifel (1957:80).

RANA MONTEZUMAE BAIRD

Rana montezumae Baird, Proc. Acad. Nat. Sci. Philadelphia, 7:61, October 20, 1854.—Mexico City, Distrito Federal, México.

La Palma; 8 km. NW of Maravatio (10); Sahuayo; Tupátaro (7).

This species probably is more abundant and widespread than is indicated by the few specimens listed above. It has been found only in the vicinity of permanent water on the Mexican Plateau and the mountains rising from the plateau at elevations of 1500 to 2000 meters. Its apparent absence from Lago de Pátzcuaro cannot be explained, unless *Rana dunni* replaces it there.

RANA PIPIENS SCHREBER

Rana pipiens Schreber, Der Naturforscher, Halle, 18:185, 1782.—Raccoon, Gloucester County, New Jersey.

Aguililla (2); Apatzingán (13); Arteaga; Axolotl (16); Camachines (2); Capirio; Cascada Tzararacua (3); Cerro San Andrés (6); Charapendo (4); Ciudad Hidalgo; Coalcomán (17); Cuitzeo (3); El Sabino (10); Jacona (3); 29 km. NW of Jacona (8); Jiquilpan; La Orilla (3); La Palma (5); La Playa (4); Lago de Chapala (3); Lago de Pátzcuaro (6); Lombardia; Los Conejos (67); Los Reyes (7); Macho de Agua; Maravatio; Morelia (5); Opopeo (3); Pátzcuaro (9); 26 km. S of Pátzcuaro (52); Puerto Hondo (3); Río Duero, 14 km. E of Zamora (13); Río Tepalcatepec, 27 km. S of Apatzingán (2); San Gregorio (38); San José de la Cumbre (5); Tangamandapio; Zacapu; 18 km. W of Zamora (35).

Except on the Pacific lowlands, this species is abundant throughout the state. It has been collected from sea level to 2800 meters, the greatest altitudinal range of any amphibian in Michoacán. It has been found frequently in the Tepalcatepec Valley; it is not a distinctly highland species in southern Michoacán, as stated by Peters (1954:9). One specimen from Aguililla (UMMZ 119257) is an albino. In this specimen there is a faint pattern on the hind limbs; otherwise the entire body is creamy white; the eyes are pink.

RANA PUSTULOSA BOULENGER

Rana pustulosa Boulenger, Ann. Mag. Nat. Hist., ser. 5, 11:343, 1883.—Ventanas, Durango, México.

Arteaga (4); 21 km. S of Arteaga; Cascada Tzararacua (3); Coalcomán (3); 12 km. ENE of Dos Aguas (3); El Sabino (53); Los Reyes (3); Tzitzio (4); Uruapan.

Although *Rana pustulosa* seems to be absent from the Mexican Plateau in Michoacán, it has been collected at elevations of 850 to 2150 meters on the slopes of the Cordillera Volcánico and in the Sierra de Coalcomán. Usually the frogs are found along rocky streams, but at Coalcomán they were found in a hyacinth-choked old river channel, and at El Sabino, in irrigation ditches.

In most specimens the dorsum is dark olive-brown; in some it is pale olive-tan with dense dark brown mottling on the back and dark transverse bands on the hind limbs.

Thirteen tadpoles (UMMZ 94271) taken from a seepage pool by a stream near Uruapan closely resemble the description of tadpoles of this species given by Taylor (1942b).

REPTILIA

TESTUDINES

CHELONIA MYDAS (LINNAEUS)

Testudo mydas Linnaeus, Systema naturae, ed. 10:197, 1758.— Type locality restricted to Ascension Island by Mertens and Müller (1928:23).

Chelonia mydas, Brongniart, Bull. Sci. Soc. Philom., 2:89, 1800.

Beach between Río Motín and Río Colotlán (2); Maruata; Playa Azul (4).

Green sea turtles are abundant along the coast of Michoacán. Laying females and fresh nests were found on August 6-12, 1950, July 14-16, 1951, and July 8-10, 1955. The general account of sea turtles on the coast of Michoacán that was given by Peters (1957) is supplemented here by my field notes on the actions of one female observed on the night of July 14, 1951, near Maruata by Donald D. Brand and I. Because of a full moon, visibility was excellent.

In the course of the day several *Chelonia* were seen in the surf; shortly after dark the first turtle was observed on the beach. Several were observed to come out on the beach and crawl nearly to the strand line, only to return to the sea.

At 10:20 p. m. one turtle was seen about 15 meters from the water. We watched this turtle from some distance and observed that by 10:26 p. m. she had moved about ten meters to a bank of sand about two meters high. Ten minutes later she had climbed the bank and disappeared over the top into the brush. We moved closer and remained hidden below the bank. Although we could not see the turtle, we could hear her movements. Between 10:37 and 10:57 p. m. the turtle dug, often flipping the dry sand for a distance of about two meters. When this energetic digging ceased, we moved up the bank to see that she was facing inland and sitting in a depression about one and one-half meters in diameter and 30 centimeters in depth. She had cleaned out this depression in the past 20 minutes. Between 11:00 and 11:36 p. m. she dug the nest hole by first scooping sand with one hind flipper and then with the other; when sand was thrown by one flipper, there was a similar, but weaker, motion by the other flipper. At 11:36 p. m. she stopped digging. By crawling up behind the turtle we were able to examine the nest cavity, which measured 21 centimeters across the top and 38 centimeters deep. The diameter of the bottom of the hole was estimated to be about 50 centimeters. At 11:40 p. m. she released the first egg; a minute later she dropped the second. At 11:42 p. m. the third and fourth eggs were released; these were coherent, as were the fifth and sixth eggs released at 11:43 p. m. After this, as many as three eggs were dropped at a time. After laying about 60 eggs, she paused for a minute and then continued laying. By 11:55 p. m. she had laid 98 eggs; after this, the process of deposition slowed considerably. She dropped a fragment of an egg followed by normal eggs. At midnight she deposited a miniature egg about 20 mm. in diameter. This terminated the deposition. Immediately she began to cover the nest.

Within ten minutes after the last egg was deposited the nest had been covered. The turtle first had been seen at 10:20 p. m.; judging from its speed and its distance from the water, the turtle probably had been on land for about ten minutes. About 25 minutes were used in crawling from the water to the nesting site. One hour and 33 minutes were spent at the nesting site; of this time twenty minutes were taken for egg deposition. The turtle was not followed back to the water, but if the return trip took approximately the same amount of time as required to travel from the ocean to the nesting site, the total elapsed time from departure to return to the water was about two and one-half hours.

We collected the eggs as they were deposited. There were 106 eggs, each having a diameter of about 40 mm., plus one small egg and a fragment of another. The turtle had a carapace about one meter in length.

From our limited observations of sea turtles and their tracks on the beaches, and from the accounts of these animals by the residents of the coastal region, great numbers of sea turtles use these relatively uninhabited beaches for nesting grounds. However, the turtles do not go unmolested. The natives capture turtles and collect their eggs. Opened and emptied nests also showed signs of predatory activity on the part of other mammals. In the vicinity of Playa Azul several turtles were killed by dogs.

KINOSTERNON HIRTIPES HIRTIPES WAGLER

Cinosternon hirtipes Wagler, Naturl. Syst. Amph., p. 37, 1830.— México. Type locality restricted to Mazatlán, Sinaloa, México, by Smith and Taylor (1950b:25).

Kinosternon hirtipes hirtipes, Schmidt, Check list N Amer. Amph. Rept., ed. 6, p. 89, 1953.

Eight km. W of Ciudad Hidalgo; Jiquilpan; La Palma; Lago de Camécuaro (4); Lago de Cuitzeo (3); Lago de Pátzcuaro (8); 14 km. E of Zamora (4).

One specimen from eight kilometers west of Ciudad Hidalgo (UIMNH 24707) is from the Río Tuxpan, a tributary of the Río Balsas; this is the only record for the species from the Balsas drainage. All others are from the lakes or rivers flowing into the lakes on the southern part of the Mexican Plateau. This species exists in Lago de Pátzcuaro to the apparent exclusion of the abundant and widespread *Kinosternon integrum*.

KINOSTERNON INTEGRUM LECONTE

Kinosternon integrum LeConte, Proc. Acad. Nat. Sci. Philadelphia, 7:183, 1854.—México. Type locality restricted to Acapulco, Guerrero, México, by Smith and Taylor (1950b:25).

Agua Cerca (3); Aguililla; Arteaga (8); Apatzingán (7); Barranca de Herradero; Buenavista (20); Capirio (2); Charapendo (3); Chupio; Coahuayana (2); Coalcomán (169); Copuyo (4); El Sabino (8); Jacona; Jiquilpan (12); La Orilla (2); La Playa (2); Lago de Cuitzeo (27); Las Higuertas; Lombardia (3); Los Reyes (5); Morelia; Ojos de Agua de San Telmo; San Pedro Naranjestila; Tacícuaro.

Excepting Lago de Pátzcuaro, *Kinosternon integrum* occupies all permanent and temporary ponds, lakes, and streams below 2200 meters throughout the state. At Coalcomán the species was in roadside ditches, small puddles, flooded fields, a hyacinth-choked ox-bow of the Río Coalcomán, as well as in the Río Coalcomán and its tributaries. Specimens from Arteaga and Barranca de Herradero were found in clear rocky streams; the one from Las Higuertas was found in a small muddy pond in pine-oak forest.

On August 26, 1960, James R. Dixon found a copulating pair in a pool at Capirio. The large series from Coalcomán contains juveniles and adults; these turtles formed the basis for the study of relative growth of plastral scutes in this species by Mosimann (1956).

GEOEMYDA RUBIDA PERIXANTHA MOSIMANN AND RABB

Geoemyda rubida perixantha Mosimann and Rabb, Occ. Pap. Mus. Zool. Univ. Michigan, 548:1, November 9, 1953.—Eight kilometers south of Tecomán, Colima, México.

Apatzingán (2); Coahuayana; La Placita; Punta San Juan de Lima.

These specimens have been discussed in detail by Mosimann and Rabb (1953). All are from the arid tropical scrub forest; those from the coastal regions were collected at elevations of less than 40 meters, and those from the Tepalcatepec Valley were collected at an elevation of 335 meters.

CROCODILIA
CROCODYLUS ACUTUS ACUTUS CUVIER

Crocodylus acutus Cuvier, Ann. Mus. Hist. Nat. Paris, 10:55, 1807.—Santo Domingo.

Crocodylus acutus acutus, Müller and Hellmich, Ibero-Amerik.

Stud., 13:128, 1940.

Boca de Apiza (2); Playa Azul (2).

The crocodile or "caiman" is abundant in the brackish lagoons along the cost of Michoacán; three large adults and several juveniles were observed at Estero Pichi at Playa Azul; others were seen at Mexiquillo and Maruata. Residents of the Balsas-Tepalcatepec Basin frequently have reported "caimanes" in the Río Balsas and Río Tepalcatepec, but the existence of the crocodile in these rivers has not been verified by specimens.

SAURIA

PHYLLODACTYLUS DUELLMANI DIXON

Phyllodactylus duellmani Dixon, Southwest Nat., 5:37, April 15, 1960. — Rancho El Espinal, Michoacán, México.

Fourteen km. SSW of Apatzingán; Capirio; Cafradía (3); El Espinal (3).

This species is known only from the Tepalcatepec Valley, where it has been found in open arid situations from 180 to 500 meters. Specimens were found in the daytime in stumps, dead cacti, and the hollow branches of the legume, *Apoplanesia paniculata*. In life adults were pale gray or grayish tan above and creamy white below. A juvenile having a snout-vent length of 18 mm. had a pale orange tail with gray cross-bands. In the adults the tail was colored like the body. The specimen from 14 kilometers south-southwest of Apatzingán (KU 29764) and those from Cofradía (BMNH 1914.1.28.28-30) were not listed by Dixon (1960).

PHYLLODACTYLUS HOMOLEPIDURUS SMITH

Phyllodactylus homolepidurus Smith, Univ. Kansas Sci. Bull., 22:121, November 15, 1935. — Five miles southwest of Hermosillo, Sonora, México.

El Ticuiz (2); La Placita; Ostula (2); Pómaro; San Pedro Naranjestila.

These specimens have been referred to *Phyllodactylus homolepidurus* by James R. Dixon (*in litt.*), who is currently studying the American members of the genus. Geckos of this species have been found in tropical semi-deciduous forest in the coastal lowlands to elevations of 500 meters. Most specimens were found beneath the bark of standing dead trees or stumps. Two individuals from El Ticuiz (UMMZ 115102) in life were dark gray above with brownish tubercles; the belly was a dusty cream-color. Apparently this species does not enter the Tepalcatepec Valley, where *Phyllodactylus lanei* is abundant.

PHYLLODACTYLUS LANEI SMITH

Phyllodactylus lanei Smith, Univ. Kansas Sci. Bull., 22:125, November 15, 1935. — Tierra Colorado, Guerrero, México.

Apatzingán (13); 21 km. S of Arteaga: El Sabino (53); La Playa; Ostula (2); Río Marquez, 10 km. S of Lombardia (8); 16 km. N of Tafetán.

This widespread species has been taken at elevations of less than 1100 meters in the Balsas-Tepalcatepec Basin, where it occurs in riparian situations in the foothills. Specimens have been collected in tropical semi-deciduous forest at Ostula and in oak forest south of Arteaga; both of these localities are on the Pacific slopes of the Sierra de Coalcomán, a region inhabited by *Phyllodactylus homolepidurus*. Both species have been collected at Ostula.

A juvenile from 21 kilometers south of Arteaga (UMMZ 118933) had alternating black and white bands on the tail. In life most of the lizards are dull ashy gray or grayish tan above and white below. According to Dixon (*in litt.*), one specimen from Apatzingán (UMMZ 115102) resembles *Phyllodactylus magnus* in scutellation, but it lacks the distinctive yellow venter of that species.

Apparently *Phyllodactylus lanei* is restricted to rather mesic environments in the Balsas-Tepalcatepec Valley and surrounding foothills; in the more open arid environments on the floor of the valley it seems to be replaced by *Phyllodactylus duellmani*.

PHYLLODACTYLUS PAUCITUBERCULATUS DIXON

Phyllodactylus paucituberculatus Dixon, Southwest. Nat., 5:40, April 15, 1960.—Río Cupatitzio (= Río Marquez), 6.5 miles south of Lombardia, Michoacán, México.

Río Marquez, 10 km. S of Lombardia (6).

Two of these specimens (UMMZ 112692-3) were discussed in detail by Dixon (1960:40) in his description of the species. On August 25, 1960, Dixon collected four additional specimens at the type locality, a conglomerate cliff along the Río Marquez. These will be reported by him in his forthcoming study of the genus.

ANOLIS DUNNI SMITH

Anolis dunni Smith, Copeia, no. 1:9, May 10, 1936.—Agua del Obispo, Guerrero, México.

Arteaga (3); 19 km. S of Arteaga.

Three females from Arteaga (UMMZ 119075) have snout-vent lengths of 41, 41, and 44 mm. In life the pale grayish brown dorsum was marked with dark brown; the belly was white, and the throat was pale pink. All have a dark interorbital bar and dark vertical bars on the upper labials. In two specimens there are only scattered dark flecks on the dorsum; in the third there is a dark postorbital stripe, a dark lateral stripe, and four narrow transverse bands on the body. A male from 19 kilometers south of Arteaga (UMMZ 119076) having a snout-vent length of 49 mm. had in life a tan dorsum, a broad white stripe from the ear to the groin, scattered small white spots on the dorsum, and indistinct pale cream-colored spots on the posterior surfaces of the thighs. This male has the dark labial bars, but lacks the dark interorbital bar, found in the females. The large rose-pink throat fan extends to about the middle of the belly. In all of the specimens the middorsal scales are keeled and much smaller than the smooth pavementlike or slightly imbricate ventrals. All have two gulars in contact with the mental, five scales between the nasals, five scales (not including the first labials) in contact with the rostral, and four rows of loreals. In these characters these specimens agree well with *Anolis dunni* from Guerrero, as diagnosed by Davis (1954b).

Previously *Anolis dunni* has been reported only from the vicinity of Agua del Obispo, Guerrero, a locality situated at an elevation of about 900 meters in pine-oak forest in the Sierra del Sur. All known close relatives of *Anolis dunni* occur only in Guerrero: *A. taylori* Smith and Spieler from Acapulco, *A. gadowi* Boulenger from Tierra Colorado, *A. liogaster* Boulenger, and *A. omiltemanus* Davis from Omiltemi. The present specimens from elevations of about 900 meters in riparian stream vegetation and oak forest represent the northern known limits of this group of *Anolis*.

ANOLIS NEBULOSUS (WIEGMANN)

Dactyloa nebulosa Wiegmann, Herpetologia Mexicana, p. 47, 1834.—México. Type locality restricted to Mazatlán, Sinaloa, México, by Smith and Taylor (1950b:66).

Anolis nebulosas, Bocourt, Mission Scientifique au Mexique et dan l'Amerique Centrale. Reptiles, livr. 2:77, 1873.

Acahuato (3); Agua Cerca; Apatzingán (4); Araparicuaro (3); 29 km. S of Ario de Rosales (3); 20 km. S of Arteaga (2); Barranca de Bejuco; Cascada Tzararacua (5); Cerro Tancítaro (13); Cherán; Chupio (5); Coalcomán (10); Cofradía; Dos Aguas (10); 18 km. E of Dos Aguas (3); El Diezmo; El Sabino (43); El Ticuiz; Jiquilpan (2); La Orilla; La Placita; La Playa (3); Los Conejos (2); Los Pozos; Nogueleras (2); Ostula; 8 km. W of Pátzcuaro (2); 8 km. NE of Pátzcuaro; Playa Azul (3); Río Cachán; Río Marquez, 10 km. S of Lombardia; Río Tepalcatepec, 27 km. S of Apatzingán; San Juan de Lima (6); San Pedro Naranjestila; Temazcal; Tuxpan (2); Tzitzio; Uruapan (74); 11 km. N of Uruapan (2); Volcán Jorullo; 16 km. E of Zacapu (2); 18 km. W of Zamora; Ziracuaretiro.

Even with the abundance of material the assignment of a specific name to these anoles is only tentative, for definite determination between *Anolis nebulosus* Wiegmann and *A. nebuloides* Bocourt is uncertain. Bocourt (1873:75) distinguished *A. nebuloides* from *A. nebulosus* by the following characters: (1) head scales keeled, not smooth; (2) snout narrower; (3) ear opening larger; (4) supraorbital semicircles separated by a row of small scales and not in contact; (5) dorsal scales larger and subequal in size to the belly scales. Boulenger (1885:77) used the same characters; Smith and Taylor (1950b:58) in their key to the Mexican species of *Anolis* stated that the dorsal scales are slightly smaller than the ventrals in *A. nebulosus* and markedly smaller in *A. nebuloides*. Smith (*in litt.*) stated that the characters of the relative sizes of the dorsal and ventral scales were incorrect in that key.

The application of the above criteria to specimens from Michoacán has not resulted in the recognition of two species. The majority of the specimens have the supraorbital semicircles separated by at least one small scale; the head scales, with the exception of those on the snout in a few individuals, are smooth; the dorsal scales are only slightly smaller than the ventrals. In other characters of scutellation the specimens are highly variable. The males in life have an orange throat fan. Anoles of this kind have been found in Michoacán, Colima, Jalisco, Nayarit, and southern Sinaloa. Near Oaxaca, Oaxaca, specimens were collected that superficially resemble those from Michoacán and farther north. These have low keels on the snout scales, dorsals somewhat larger than the ventrals, and a pink throat fan. In ten males from Oaxaca the size of the dorsal scales relative to that of the ventrals is 1.00:0.83; the same ratio for 25 males from Michoacán is 1.00:1.08. In both samples there are specimens in which the dorsal and ventral scales are about equal in size.

Investigations by Richard E. Etheridge on the osteology of *Anolis*, including those species here being considered, have revealed relatively constant differences in the parasternalia and in the caudal vertebrae. The application of Etheridge's findings to anoline systematics must await the completion of his study.

The carination of the scales on the snout *versus* smooth scales there seems to be the only significant character given by Bocourt that distinguishes *A. Nebuloides* from *A. nebulosus*. The difference in the color of the throat fan, which is apparent only in living individuals, is more striking. Obviously more than one species is represented, as is borne out by the differences in the color of the throat fan and in the osteology, but there is uncertainty about the correct name for each species. On the strength of Bocourt's diagnosis of keeled snout scales in *A. nebuloides*, I am applying that name to the population in Oaxaca and *A. nebulosus* to the specimens from Michoacán. As arranged here, the two species can be distinguished, as follows:

> *A. nebulosus.*—Dorsal scales only slightly smaller than the ventral scales; snout scales usually smooth; throat-fan bright orange in adult males.

> *A. nebuloides.*—Dorsal scales somewhat larger than the ventral scales; snout scales having a low keel; throat-fan pink in adult males.

With respect to geographic distribution, *A. nebulosus* has been collected from southern Sinaloa southward to Michoacán. The lizards here referred to *A. nebuloides* have been taken only in pine-oak forest on the mountain slopes near Oaxaca City. Zweifel and Norris (1955:233) reported anoles with pink throat-fans from southern Sonora; possibly those specimens are *A. nebuloides*; I have not examined them. I have seen several preserved specimens from the vicinity of Tehuantepec, Oaxaca. Although they probably belong to this group, those specimens differ from both *A. nebulosus* and *A. nebuloides* in their larger size, relatively larger head, and much larger throat fan.

Aside from the minor variation in scutellation, specimens of *Anolis nebulosus* from Michoacán vary greatly in coloration. Usually the females have some form of a broad middorsal pale-colored band. In life this is dull yellow, tan, or orange. Two females from Dos Aguas are strikingly different; one (UMMZ 119521) has a broad middorsal orange stripe that is scalloped laterally and bordered by gray. The other (UMMZ 119081) has a narrow middorsal cream-colored line. Males usually are unicolor brown or olive-tan; sometimes the middorsal region is darker. Some individuals have dark cross-bands or chevrons on the dorsum. One male from Dos Aguas (UMMZ 119080) has a cream-colored lateral stripe.

In Michoacán *Anolis nebulosus* occurs from sea level to elevations slightly in excess of 2100 meters, usually in areas of dense cover, whether this be herbaceous, viney, or woody, ordinarily on the ground as well as in bushes and trees. One was in a bromeliad growing about ten meters above the ground. In the arid Tepalcatepec Valley anoles of this species are most frequently found in the tangled growth along streams. Above Uruapan they were found in pine-oak forest, and on the Mexican Plateau between Zamora and Zacapu they were found in a bunch grass-scrub oak association.

ANOLIS SCHMIDTI SMITH

Anolis schmidti Smith, Publ. Field Mus. Nat. Hist., zool. ser., 24:21, January 30, 1939.—Manzanillo, Colima, México.

La Placita; San Juan de Lima.

Peters (1954:11) reported on the specimen from La Placita; another was secured at San Juan de Lima in 1956. The latter (UMMZ 115078) is a male having a snout-vent length of 43.0 mm. and a tail length of 70.5 mm. The dorsal ground color is pale tan; there are five pairs of irregular dark brown dorsolateral blotches. In life the throat fan was pale orange. These specimens agree with those from Colima described by Duellman (1958c:10). The distribution of *Anolis schmidti* seems to be restricted to the coastal lowlands from Michoacán to Nayarit.

BASILISCUS VITTATUS WIEGMANN

Basiliscus vittatus Wiegmann, Isis von Oken, 21:373, 1828.— México. Type locality restricted to Veracruz, Veracruz, México, by Smith and Taylor (1950b:72).

Apatzingán (9); Capirio; Coahuayana (5); El Cerrito; El Sabino (2); El Ticuiz; La Placita (3); Maruata (2); Motín del Oro; Ostula; Playa Azul (3).

This species has been found only on the coast and in the low Tepalcatepec Valley. In the latter area it is restricted to riparian situations along the larger streams. The lizard is abundant in the mangrove swamps bordering the brackish lagoons on the coast. In July, 1955, scores of individuals were seen around Estero Pichi at Playa Azul. Adults, especially the large males, are exceedingly wary and difficult to collect. At all localities where they were found, the lizards were most often seen in dense bushes, where they are well camouflaged. Individuals of all sizes were observed to run across the surface of the ponds.

IGUANA IGUANA RHINOLOPHA WIEGMANN

Iguana rhinolopha Wiegmann, Herpetologia Mexicana, p. 44, 1834.—México. Type locality restricted to Córdoba, Veracruz, México, by Smith and Taylor (1950b:72).

Iguana iguana rhinolopha, Van Denburgh, Proc. Acad. Nat. Sci. Philadelphia, 1897:461, January 18, 1898.

Apatzingán (8); Capirio (3); El Cerrito; El Ticuiz (2); La Placita; La Playa (2); Maruata; Playa Azul; Río Cachán.

Like the preceding species, this lizard is always found near water. It does not ascend the foothills of the Sierra de Coalcomán, but in the Balsas Basin it reaches elevations of 800 meters at La Playa. Large adults are often seen in the large trees making up the gallery forests along rivers. From high perches the lizards drop into the water with a terrific splash. Bright green juveniles were abundant in bushes along the Río Tepalcatepec in July, 1955.

CTENOSAURA PECTINATA (WIEGMANN)

Cyclura pectinata Wiegmann, Herpetologia Mexicana, p. 42, 1834.—México (by inference). Type locality restricted to Colima, Colima, México, by Bailey (1928:25).

Ctenosaura pectinata, Gray, Catalogue of the lizards... British Museum, p. 191, 1845.

Apatzingán (27); between Ario de Rosales and La Playa; Barranca de Bejuco; Capirio (2); Coalcomán (4); El Espinal; El Sabino (2); El Ticuiz; Jazmin (2); La Huacana; La Placita (8); La Playa (3); Limoncito; Lombardia; Motín del Oro; Playa Azul; Río Cancita, 12 km. E of Apatzingán (2); Río Marquez, 10 km. S of Lombardia (2);? Uruapan; Volcán Jorullo.

Ctenosaura pectinata is a common lowland species that ascends the slopes of the Sierra de Coalcomán and the Cordillera Volcánica to elevations of about 1050 meters (approximating the lower limits of the oak forest). The record from Uruapan (USNM 10234, collected by Dugès) is doubtful.

These large lizards are most easily observed on rock fences along roads. Near Apatzingán innumerable individuals can be seen in mid-morning. Later in the day, as the sun rises higher in the sky, the lizards retreat to the shade of the crevices in the fences. The abundance of these lizards in the Tepalcatepec Valley, together with evidence gathered from the natives of the valley, indicates that these lizards are seldom used for human consumption there. On the other hand, several people in Coalcomán consider the "iguana negra" (local name for *Ctenosaura*) to be a delicacy and serve it at every opportunity. In early July, 1951, brilliant green young of the year were collected at La Playa and at Coalcomán.

ENYALIOSAURUS CLARKI (BAILEY)

Ctenosaura clarki Bailey, Proc. U. S. Natl. Mus., 73:44, September 26, 1928.—Ovopeo (= Oropeo), Michoacán, México.

Enyaliosaurus clarki, Duellman and Duellman, Occ. Pap. Mus. Zool. Univ. Michigan, 598:1, February 16, 1959.

Twelve km. SSW of Apatzingán; Capirio (7); Cofradía (3); El Espinal (2); 32 km. E of Huetamo; Jazmin (5); Oropeo (10); Rancho Nuevo; Río Cancita, 12 km. E of Apatzingán (8); Tepalcatepec (3); Zicuiran (6).

This species is known only from the low areas of the Balsas-Tepalcatepec Basin between elevations of 200 and 510 meters. It is commonly found in the open arid tropical scrub forest dominated by *Prosopsis* sp., *Apoplanesia paniculata*, and *Cercidium plurifoliolatum*. Continued collecting in the Tepalcatepec Valley has borne out the suggestions of Duellman and Duellman (1959) concerning the distribution and abundance of this lizard. Also, continued collecting in Colima and on the Pacific coast has failed to reveal the presence of *Enyaliosaurus* there.

PHRYNOSOMA ASIO COPE

Phrynosoma asio Cope, Proc. Acad. Nat. Sci. Philadelphia, 16:178, September 30, 1864.—Colima, Colima, México.

Apatzingán (4); San Salvador.

In Michoacán this species has been obtained only in the Tepalcatepec Valley and on the northern slopes of the Sierra de Coalcomán between 300 and 700 meters. Apparently the lizard is absent from the coastal lowlands of Michoacán and Guerrero. The distribution of this species, therefore, is discontinuous. One population inhabits the lowlands of Colima and the Balsas-Tepalcatepec Basin inland to northern Guerrero and Morelos; a southern population inhabits the Plains of Tehuantepec in Oaxaca.

A juvenile from Apatzingán (USNM 47739) has a snout-vent length of 40.0 mm. and a tail length of 19.5 mm.

SCELOPORUS AENEUS AENEUS WIEGMANN

Sceloporus aeneus Wiegmann, Isis von Oken, 21:370, 1828.— México. Type locality restricted to Tres Cumbres, Morelos, México, by Smith and Taylor (1950b:137).

Sceloporus aeneus aeneus, Smith, Occ. Pap. Mus. Zool. Univ. Michigan, 361:6, December 15, 1937.

Angahuan; Araparicuaro (2); Capácuaro (2); Carapan (11); Cherán (11); 18 km. WNW of Ciudad Hidalgo (10); Cuseño Station; Jeráhuaro; Los Conejos (36); Macho de Agua (7); Opopeo; Paracho (2); Pátzcuaro (4); Pino Gordo; 18 km. W of Quiroga (2); Tancítaro (49); Uruapan (14); 16 km. NW of Zacapu (5); between Zacapu and Zamora (2); 13 km. E of Zinapécuaro; 14 km. SE of Zitácuaro (14).

This small terrestrial species inhabits the pine and fir forests of the Cordillera Volcánica between elevations of 1850 and 3100 meters; apparently it is absent from the Sierra de Coalcomán. It seems to prefer rather open coniferous forests in which there is a more or less continuous cover of grasses on the ground. On warm sunny days the lizards can be observed scurrying about in the grass; in the early hours of the day, or on cold days, they are found beneath stones, logs, or dead clumps of bunch grass.

SCELOPORUS ASPER BOULENGER

Sceloporus asper Boulenger, Proc. Zool. Soc. London, 1897:497, October, 1897.—La Cumbre de los Arrastrados, Jalisco, México.

Apatzingán (3); 10 km. E of Dos Aguas; Uruapan (41).

This strictly arboreal lizard is abundant in the mixed broad-leafed forest near Uruapan. The lizards are exceedingly wary and can be approached only with difficulty. In life males have pale blue bellies; the throat is pale pink. The pale gray dorsum marked with irregular darker gray blotches blends well with the color of the tree trunks on which the lizard lives. The one specimen from Dos Aguas was found on a pine tree; it provides the only record for the species from the Sierra de Coalcomán.

SCELOPORUS BULLERI BOULENGER

Sceloporus bulleri Boulenger, Proc. Zool. Soc. London, 1894:729, April, 1895.—Las Cumbre de los Arrastrados, Jalisco, México.

Acuaro de las Lleguas (13); Barolosa (9); Dos Aguas (61); 10 km. NE of Dos Aguas (5).

Heretofore this species has been known only from a few specimens from scattered localities in the Sierra Madre Occidental in southwestern Jalisco and Sinaloa. The collection of a large series of these lizards in virgin pine forest at elevations of more than 2000 meters in the Sierra de Coalcomán now makes possible an analysis of variation in the species.

Superficially *S. bulleri* resembles *S. torquatus,* but *S. bulleri* is smaller, has more dorsal scales, fewer scales in the dark collar, and fewer femoral pores. In 88 specimens of *S. bulleri* there are 36-41 (38.7) dorsal scales and 2 or 3 (2.6) middorsal scales in the collar, as compared with 28-31 (29.3) dorsal scales and 3 or 4 (3.4) middorsal scales in the collar of 26 specimens of *S. torquatus* from Uruapan. In 20 adult males of *S. bulleri* there are 13-15 (14.3) femoral pores, and 13-16 (14.4) in 11 females; 13 males of *S. torquatus* have 14-21 (17.3) femoral pores, and 13 females have 15-21 (16.7). Seventeen adult males of *S. bulleri* have snout-vent lengths of 72-91 (82.0); ten females, 71-87 (75.7). In comparison, 13 adult males of *S. torquatus* have an average snout-vent length of 88.9 mm., and 13 females, 88.5 mm. In *S. bulleri* there is little variation in the head scales. The frontal is in contact with the interparietal in 63, and not in 24, specimens; the median frontonasal is in contact with the frontal in 13, and not in 74, specimens. In 39 specimens there are two canthals, and in 48 there is one; in 29 specimens there are three preauriculars, and in 58 there are four.

In life adult males have a pale blue tail, bright blue belly patches, a purplish blue throat, and pale blue lines on the sides of the head and neck.

This species was obtained at four localities in the high mountains of the Sierra de Coalcomán. In this mountain range *Sceloporus bulleri* apparently replaces *S. torquatus*, a species that is widespread in the Cordillera Volcánica and on the Mexican Plateau. At Dos Aguas and at Acuaro de las Lleguas the lizards were abundant in the tall pine forest, where they were found on standing pine trees, on pine logs, and on rock outcroppings.

SCELOPORUS DUGESI INTERMEDIUS DUGÈS

Sceloporus intermedius Dugès, La Naturaleza, 4:29, 1877.—La Noria, near Zamora, Michoacán, México.

Sceloporus dugesii intermedius, Smith, Univ. Kansas Sci. Bull., 24:663, February 16, 1938.

Cojumatlán (6); Jiquilpan (11); Lago de Camécuaro; Lago de Chapala; Morelia (23); Pátzcuaro (84); Quiroga (35); Sahuayo (4); Tacícuaro (2); Tangamandapio (17); Tangancícuaro (9); Zacapu (4); Zamora (11); Zinapécuaro (9).

This lizard is strictly an inhabitant of the Mexican Plateau, where it is found in rocky places, sometimes in pine-oak forest, but more frequently in mesquite-grassland. It is a terrestrial species, and is most often seen on rock fences at elevations of 1500 to 2200 meters.

This species differs from *S. bulleri* and *S. torquatus* in having two rows of supraoculars, instead of one; also it has more dorsal scales. Twenty-six specimens of *Sceloporus dugesi intermedius* from Tangamandapio and Tangancícuaro have 44-48 (45.7) dorsal scales, as compared with an average of 38.7 in *S. bulleri* and 29.3 in *S. torquatus*. In life *Sceloporus dugesi intermedius* has a dull greenish gray dorsum; in males the belly patches are bright blue bordered medially by black, and the throat is bluish gray. The largest specimen examined is a male having a snout-vent length of 80 mm.

SCELOPORUS GADOWAE BOULENGER

Sceloporus gadoviae Boulenger, Proc. Zool. Soc. London, 1905, 2:246, October 7, 1905.—Mezquititlán, Guerrero, México.

Chupio; El Sabino (77); La Playa (6); Río Marquez, 10 km. S of Lombardia (11).

Although this species has a rather extensive range in the Balsas-Tepalcatepec Basin in the state of Michoacán, Guerrero, Morelos, and Puebla, it is only locally abundant in that area. Usually these lizards are found on rocky cliffs in which there are many crevices for cover. *Sceloporus gadowae* is abundant on a conglomerate cliff along the Río Marquez south of Lombardia. Although the closely related *S. pyrocephalus* is abundant in the stream valley and in the hills above the cliff, *S.*

gadowae has been found only on the cliff; few individuals of *S. pyrocephalus* have been observed on the cliff. A similar situation was discovered on a much more extensive conglomerate cliff along the Río Balsas near Mexcala, Guerrero. Near Tehuitzingo, Puebla, where *S. pyrocephalus* was not found, *S. gadowae* was found on conglomerate cliffs. Probably there is strong competition between the two species; possibly this has resulted in the restriction of *S. gadowae* to isolated cliff-habitats within the extensive range of the more widespread *S. pyrocephalus*.

In Michoacán *Sceloporus gadowae* has been found along the lower slopes of the Cordillera Volcánica at elevations from 250 to 1050 meters. All of the localities from which this lizard is known lie in the arid tropical scrub forest.

SCELOPORUS GRAMMICUS MICROLEPIDOTUS WIEGMANN

Sceloporus microlepidotus Wiegmann, Herpetologia Mexicana, p. 51, 1834. — México. Type locality restricted to México, Distrito Federal, by Smith and Taylor (1950b:120).

Sceloporus grammicus microlepidotus, Smith and Laufe, Trans. Kansas Acad. Sci., 48:332, December, 1945.

Angahuari; Apo (10); Atzimba (3); Carapan (5); Cerro San Andrés (17); Cerro Tancítaro (18); Corupu; Cuseño Station (2); Jacona; Jeráhuaro (10); Macho de Agua; Mil Cumbres; 46 km. E of Morelia; 60 km. E of Morelia (2); Opopeo (14); Pátzcuaro (30); Puerto Hondo (19); San Gregorio (41); San José de la Cumbre (8); Sierra Patamba; Tancítaro (233); Tupátaro; Undameo; Uruapan (180); between Zacapu and Zamora; 24 km. SE of Zitácuaro; between Zurumbeneo and Cerro Garnica.

This small species of *Sceloporus* is an ubiquitous inhabitant of the coniferous forests from 1550 to 3100 meters in the Cordillera Volcánica. Usually it is seen on tree trunks, but occasionally on the ground. Near the lower limit of the altitudinal distribution of the species, as at Uruapan, individuals sometimes are found on broad-leafed trees. Apparently *Sceloporus heterolepis* replaces *S. grammicus microlepidotus* in the Sierra de Coalcomán.

SCELOPORUS HETEROLEPIS BOULENGER

Sceloporus heterolepis Boulenger, Proc. Zool. Soc. London, 1894:731, April, 1895. — La Cumbre de los Arrastrados, Jalisco, México.

Araparicuaro; Cerro Barolosa (6); Dos Aguas (13); Los Conejos; 11 km. N of Uruapan (3).

Although Michoacán has not previously been included in the range of this lizard, it was first collected in the state by Gadow in 1908 (BMNH 1914.1.28.69 from Araparicuaro). The description of *S. heterolepis* given by Smith (1939:197) can be supplemented by data on the 23 specimens now in the collections of the Museum of Zoology at the University of Michigan. All have two canthals; there are 55 to 71

(63.6) scales in the middorsal row; 1 to 3 rows middorsally are somewhat enlarged and bordered on either side by a row of larger scales bearing high keels. There are 14 to 20 (16.2) femoral pores. Eight adult males have snout-vent lengths from 49 to 61 (58.0) mm. and tail lengths from 57 to 74 (66.0) mm.; four adult females have snout-vent lengths from 52 to 57 (55.2) mm. and tail lengths from 60 to 66 (63.5) mm. The smallest of eight juveniles has a snout-vent length of 28 mm. and a tail length of 32 mm. The dorsum in adults is pale grayish brown; there are three irregular chevron-shaped dark marks and a triangular dark brown mark above the insertion of the hind limbs; on the tail are dark brown rings. There are scattered faint blue flecks on the flanks and narrow transverse dark lines on the lower limbs. Males have pale bluish green belly patches and an orange-salmon-colored throat; the belly in females is pale orange-tan. The juveniles have a more contrasting color pattern; the dark chevrons on the dorsum are bordered posteriorly by pale gray.

In Michoacán this species has been obtained in pine and pine-fir forests from 1800 to 2700 meters. On Cerro Barolosa and at Dos Aguas, both in the Sierra de Coalcomán, the lizards were found beneath the bark of dead, standing pines. In the Sierra de Coalcomán *Sceloporus heterolepis* seems to fill the niche of the small arboreal *Sceloporus* in the coniferous forest in southwestern México, a position held by *S. grammicus microlepidotus* in the Cordillera Volcánica; the latter species does not occur in the Sierra de Coalcomán. Five specimens of *Sceloporus heterolepis* are known from the Cordillera Volcánica, whereas 603 of *S. grammicus microlepidotus* have been collected there. The ecological relationships that exist between the two species in the Cordillera Volcánica are not known.

Insofar as is known, *Sceloporus heterolepis* reaches the southern limits of its range in the Sierra de Coalcomán and in the western part of the Cordillera Volcánica. Other records for the species are from the Sierra Madre Occidental in Jalisco. Langebartel (1959) described *Sceloporus shannonorum* from the mountains near the Durango-Sinaloa border; the single specimen of *S. shannonorum* differs significantly from *S. heterolepis* only in having fewer dorsal scales (48). The acquisition of additional material, especially from Nayarit and northern Jalisco, probably will provide a basis for showing that these two populations are conspecific.

SCELOPORUS HORRIDUS OLIGOPORUS COPE

Sceloporus oligoporus Cope, Proc. Acad. Nat. Sci. Philadelphia, 16:177, September 30, 1864.—Colima, Colima, México.

Sceloporus horridus oligoporus, Taylor, Univ. Kansas Sci. Bull., 24:520, February 16, 1938.

Aguililla; Apatzingán (50); Arteaga (2); Capirio (2); Cascada Tzararacua; Charapendo (4); Coahuayana (3); Coalcomán (32); 19 km. S of Corralito; 27 km. E of Dos Aguas; El Sabino (55); El Ticuiz; Huetamo (2); Jazmin; Jungapeo (2); La Orilla (2); La Placita; Limoncito (3); Playa Azul (5); Tzitzio (8); Uruapan (4); Volcán Jorullo (2); Ziracuaretiro; Zirimícuaro (13).

All of the specimens from Michoacán seem to be typical *S. horridus oligoporus;*

none has more than six femoral pores.

Characteristically this species is found in open arid scrub forest; it reaches its greatest abundance in rocky areas in which there are scattered leguminous trees and bushes. It has been found in these low trees and bushes almost as frequently as it has been found on the ground; none has been seen in large trees or far above the ground. Altitudinally, this species ranges from sea level to about 1600 meters.

SCELOPORUS MELANORHINUS CALLIGASTER SMITH

Sceloporus melanorhinus calligaster Smith, Proc. U. S. Natl. Mus., 92:360, November 5, 1942.—Acapulco, Guerrero, México.

Aguililla; Apatzingán (18); Barranca de Herradero; Capirio (19); Coahuayana (4); Coalcomán (2); Cofradía (4); El Cerrito; El Sabino (33); El Ticuiz (3); La Placita (6); Lombardia (4); Playa Azul; Río Marquez, 10 km. S of Lombardia (2); Río Marquez, 13 km. SE of Nueva Italia (4); Salitre de Estopila; San Juan de Lima (2); Santa Ana; Tzitzio; Ziracuaretiro.

Smith (1942a:360) diagnosed *Sceloporus melanorhinus calligaster* as having fewer femoral pores than the other subspecies of *S. melanorhinus* and as having the lateral belly patches in the males confluent in the midline. Examination of forty specimens from the Tepalcatepec Valley and the coastal regions of Michoacán does not substantiate this diagnosis. The number of femoral pores varies from 15 to 22 (18.9); 14 individuals (35%) had 20 or more femoral pores. Smith (*loc. cit.*) stated that *S. melanorhinus* in Oaxaca had 18 to 27 (21.6) femoral pores and that 77 per cent of the specimens had more than 20 femoral pores. Of the 24 males examined from Michoacán, 18 have the lateral belly patches separated in the midline. Usually they are separated by no more than one scale, but in some individuals they are separated by two or more scales. Although the above data minimize certain differences between the northern and southern populations of this species, certain of the color pattern characters seem to be diagnostic of the subspecies inhabiting the Pacific lowlands from Guerrero to Nayarit. Large adults of *S. m. calligaster* have only a faint dorsal pattern, which in the subspecies *melanorhinus* and *stuarti* consists of a series of large, dark, interconnected triangles on the back. This pattern is present in young and small adults of *S. m. calligaster*; furthermore, in this subspecies the ventral coloration of the males differs from that found in the more southern populations. Adult males of *S. m. calligaster* have a black throat, that changes to brilliant blue posteriorly, and a large white spot medially on the chin. This spot is present in some specimens from Oaxaca and Chiapas, but, if present, it is much smaller and less distinct than in specimens from Michoacán. In *S. m. calligaster* the chest and midventral area are orange to salmon-color.

A male from Lombardia in life was colored as follows: Dorsum grayish tan bearing faint bluish gray flecks; chest deep salmon-orange, this color continuing down midventral area to the somewhat paler groin; belly patches pale blue fading to pale green laterally; throat black anteriorly enclosing a white spot; throat blue posteriorly and bluish green posterolaterally.

Individual lizards were observed to change in dorsal color from a pale ashy gray to a rather dull brown. Normally, inactive individuals and those observed on overcast days were dull brown.

Sceloporus melanorhinus calligaster is found in trees in riparian situations in the lowlands to elevations of about 1500 meters. It does not inhabit the arid tropical scrub forest in the Tepalcatepec Valley or on the coast, but in those areas is found in the gallery forests along streams and rivers. The lizards are wary and live high in the trees; they are especially difficult to locate in the rainy season, when the trees are in full leaf.

SCELOPORUS PYROCEPHALUS COPE

Sceloporus pyrocephalus Cope, Proc. Acad. Nat. Sci. Philadelphia, 16:177, September 30, 1864.—Colima, Colima, México.

Acahuato (2); Apatzingán (142); Arteaga (4); 26 km. S of Arteaga (4); Capirio (6); Chinapa; Chupio; 19 km. S of Corralito (5); El Sabino (220); Jazmin (3); La Placita (8); La Playa (14); La Salada (6); Lombardia (5); Nueva Italia (14); Ojos de Agua de San Telmo (2); Oropeo (3); Ostula; Punta de San Telmo (3); Río Cancita, 14 km. E of Apatzingán (13); Río Marquez, 10 km. S of Lombardia (10); Río Marquez, 13 km. SE of Nueva Italia (3); San Juan de Lima (2); Santa Ana (2); Tafetan (2); Tepalcatepec (2); Tzitzio (6); Volcán Jorullo (3).

This small species is extremely common in the Tepalcatepec Valley and noticeably less so on the coast. It is usually found on the ground in rocky areas, but males frequently have been seen on the trunks of low trees in the scrub forest. Altitudinally, it ranges from sea level to slightly more than 1000 meters. The sexes are readily distinguished in the field (Oliver, 1937; Smith, 1939; Duellman, 1954b). In the dry season only males were observed in the Tepalcatepec Valley, but in the rainy season both sexes were found in approximately the same numbers.

SCELOPORUS SCALARIS SCALARIS WIEGMANN

Sceloporus scalaris Wiegmann, Isis von Oken, 21:370, 1828.—México. Type locality restricted to México, Distrito Federal, by Smith and Taylor (1950b:137).

Sceloporus scalaris scalaris, Smith, Occ. Pap. Mus. Zool. Univ. Michigan, 361:2, December 15, 1937.

Carapan (2); Cherán; Ciudad Hidalgo; Huingo (3); Jacona (3); Jiquilpan (2); Lago de Camécuaro (2); Lago de Chapala; Lago de Cuitzeo (5); Morelia (4); Pátzcuaro (4); Queréndaro; Quiroga; Tacícuaro (5); Tarécuaro; Zacapu (4); Zamora (4); Zinapécuaro (11).

This small terrestrial species does not seem to be abundant anywhere in the state. It sometimes is found in open pine, oak, or pine-oak forest, but usually it is observed in areas supporting bunch grass. In such places the lizards sun and forage on the open ground and quickly take refuge in the large clumps of grass. Altitudinally, the species ranges from 1550 to 2300 meters. Although *Sceloporus scalaris scalaris* has been found in association with *S. dugesi intermedius*, *S. spinosus*, and *S. torquatus*, it does not seem to form any close ecological association with any of these species. In the pine forests of the Cordillera Volcánica *S. s. scalaris* is replaced by *Sceloporus aeneus aeneus*, another small terrestrial species that occurs in great abundance throughout the coniferous forests of the Cordillera Volcánica.

SCELOPORUS SINIFERUS SINIFERUS COPE

Sceloporus siniferus Cope, Proc. Amer. Philos. Soc., 11:159, 1869.—Pacific side of the Isthmus of Tehuantepec. Type locality restricted to Tehuantepec, Oaxaca, México, by Smith and Taylor (1950b:134).

Sceloporus siniferus siniferus, Smith and Taylor, Bull. U. S. Natl. Mus., 199:134, October 26, 1950.

Twenty-six km. S of Arteaga; Barranca de Bejuco (2); Coahuayana; El Ticuiz (2); La Mira; La Orilla (2); La Placita (9); Maruata; Ojos de Agua de San Telmo; Ostula (4); Playa Azul (6); Pómaro (2); Puerto de las Higuerita; Santa Ana (3).

This small terrestrial species inhabits the dense arid tropical scrub forest on the coast and lower foothills of the Sierra de Coalcomán to elevations of about 150 meters. It also occurs in the lower Balsas Valley, but it has not been found in the scrub forest of the broad Tepalcatepec Valley. Perhaps the large number of *Sceloporus siniferus* on the coastal lowlands is responsible for the small number there of *S. pyrocephalus*, another terrestrial species of about the same size. The latter is abundant in the Tepalcatepec Valley, where *S. siniferus siniferus* has not been found. *Sceloporus siniferus siniferus* is a fast runner and difficult to collect; consequently, the small number of specimens available is not indicative of its abundance.

SCELOPORUS SPINOSUS SPINOSUS WIEGMANN

Sceloporus spinosus Wiegmann, Isis von Oken, 21:370, 1828.— México. Type locality restricted to Puebla, Puebla, México, by Smith and Taylor (1950b:116).

Sceloporus spinosus spinosus, Martín del Campo, Anal. Inst. Biol. México, 8:262, 1937.

Cojumatlán (2); Huetamo Road; Lago de Cuitzeo (4); Maravatio (8); Tupátaro (2).

Although this species is widespread on the southern part of the Mexican Plateau, it is uncommon in Michoacán. It has been collected only in rather open situations in the mesquite-grassland on the plateau between 1500 and 2300 meters, where it has been found in association with *Sceloporus dugesi intermedius* and *S. scalaris scalaris*. Most specimens of *Sceloporus spinosus spinosus* have been observed on rock fences. In this habitat the species is the larger member of a pair of species, the smaller of which is *Sceloporus dugesi intermedius*.

SCELOPORUS TORQUATUS TORQUATUS WIEGMANN

Sceloporus torquatus Wiegmann, Isis von Oken, 21:369, 1828.—México. Type locality restricted to México, Distrito Federal, by Smith and Taylor (1950b:126).

Sceloporus torquatus torquatus, Cope, Proc. Amer. Philos. Soc., 22:402, 1885.

Angahuan (31); Araparicuaro; Capácuaro (3); Carapan (11); Cerro Tancítaro; Cherán; Ciudad Hidalgo; Cojumatlán; Copándaro (2); Corupu (4); Cuseño Station (9); El Álamo; Jacona (6); Jiquilpan (2); Jungapeo (3); Lago de Camécuaro; Lago de Chapala; Lago de Cuitzeo (3); La Palma (2); Los Conejos (3); Los Reyes (3); Maravatio (9); Morelia (17); Paracho (3); Pátzcuaro (27); Pino Gordo; Queréndaro (2); Quiroga; Sahuayo (3); San José de la Cumbre; San Juan de Panangaricutiro; Tacícuaro (10); Tancítaro (200); Tangamandapio; Tangancícuaro (3); Temazcal (2); Tupátaro (5); Uruapan (136); Zacapu; Zinapécuaro (10); Zirimícuaro (12); Zitácuaro.

This large species inhabits the Mexican Plateau and the Cordillera Volcánica, but not the Sierra de Coalcomán, where apparently it is replaced by *Sceloporus bulleri*. *Sceloporus torquatus torquatus* usually is found in pine or pine-fir forests at elevations between 1450 and 3000 meters. In many places it is almost entirely arboreal, but in areas where there are many fallen trees or rock fences and rock piles, many individuals have been found on the ground near the rocks or logs. In the coniferous forests this species is associated with *S. grammicus microtepidotus* and *S. aeneus aeneus*.

The distinction made by Smith (1938:572) between the subspecies *S. torquatus torquatus* and *melanogaster* is slight. Individuals with pale bluish spots are found throughout the range of the species in Michoacán; spotting is especially evident in the young. Individuals having an incomplete nuchal collar have been found at Maravatio and at Zinapécuaro in the northern part of the state; in this character these specimens resemble *S. torquatus melanogaster*, which is found to the north from Guanajuato to Zacatecas and San Luis Potosí.

SCELOPORUS UTIFORMIS COPE

Sceloporus utiformis Cope, Proc. Acad. Nat. Sci. Philadelphia, 16:177, September 30, 1864.—Colima, Colima, México.

Nineteen km. S of Arteaga (2); Cascada Tzararacua (17); Coahuayana (3); Coalcomán (6); El Sabino (2); El Ticuiz (2); Ostula (3); Pómaro; Río Cachán; San Juan de Lima; Uruapan (26).

In Michoacán the range of this species is discontinuous. It has been found between 1050 and 1550 meters on the slopes of the Cordillera Volcánica, and on the coast and seaward slopes of the Sierra de Coalcomán up to an elevation of 900 meters. It is absent from the Tepalcatepec Valley. At Uruapan and at Cascada Tzararacua this lizard was found on the ground in oak forest or in open pine-oak forest; on the coast and foothills of the Sierra de Coalcomán it was found on the ground in the gallery forests along streams, and not in the scrub forest.

UROSAURUS BICARINATUS TUBERCULATUS (SCHMIDT)

Uta tuberculata Schmidt, Amer. Mus. Novitates, 22:4, December 1, 1921.—Colima, Colima, México.

Urosaurus bicarinatus tuberculatus, Mittleman, Bull. Mus. Comp. Zool., 91:169, September, 1942.

Twenty-six km. S of Arteaga; Cascada Tzararacua (2); Chupio; Coahuayana; Coalcomán (8); El Sabino (2); Jungapeo; La Orilla (2); La Placita (4); Playa Azul (4); Pómaro (2); San Salvador (16);? Tupátaro; Uruapan (12); Tzitzio; Zamora.

The known distribution and geographic variation of *Urosaurus bicarinatus* in southwestern México presents a confused picture. In general rugosity, specimens from the coastal region of Michoacán (Coahuayana, La Orilla, La Placita, Playa Azul, and Pómaro) resemble *U. bicarinatus tuberculatus* to the north along the Pacific coast. Furthermore, specimens from the coast have less stippling in the gular region than do those from the Sierra de Coalcomán and the slopes of the Cordillera Volcánica. Specimens from the mountains have greatly carinate enlarged dorsals, large lateral tubercles, and heavily stippled throats; in these characters they resemble specimens from Morelos, Guerrero, and Oaxaca. As mentioned by Peters (1954:14), some specimens from La Orilla and San Salvador are like *U. bicarinatus bicarinatus* in certain characters, and one specimen has the blue ventral patches restricted to the sternal area, a characteristic of *U. bicarinatus anonymorphus* of Oaxaca and eastern Guerrero. Examination of all available specimens from Michoacán indicates that the nature of the dorsal scales is of little value in separating the subspecies. The specimens from Michoacán are here provisionally referred to *U. bicarinatus tuberculatus*, because cursory examination of specimens from several localities between Nayarit and Oaxaca shows that there are only minor differences between the named populations. Individuals from the northern part of the range are more rugose and have larger blue ventral patches and less gular stippling than those from the south.

In Michoacán *Urosaurus bicarinatus tuberculatus* is found in wooded areas, not in open scrub forest, in the coastal area to elevations of about 900 meters, and along the slopes of the Cordillera Volcánica and the southern edge of the Mexican Plateau at elevations from 1000 to 1700 meters. The record for Tupátaro probably is erroneous, for no other specimens of this species are known from the central plateau. Essentially, the distribution of this species parallels that of *Sceloporus utiformis*, a strictly terrestrial species. *Urosaurus bicarinatus tuberculatus* lives on tree trunks. Below 1000 meters in the Tepalcatepec Valley *Urosaurus bicarinatus tuberculatus* is replaced by *Urosaurus gadowi*.

UROSAURUS GADOWI (SCHMIDT)

Uta gadovi Schmidt, Amer. Mus. Novitates, 22:3, December 1, 1921.—Cofradía, Jalisco, México (in error) = Cofradía, Michoacán, México (Duellman, 1958b:49).

Urosaurus gadowi, Mittleman, Bull. Mus. Comp. Zool., 91:154, September, 1942.

Acahuato (2); Apatzingán (56); 12-16 km. S of Apatzingán (12); Buenavista (7); Capirio (23); Cofradía (21); El Sabino (13); Guayabo; Jazmin; La Playa; La Salada (3); Nueva Italia (7); Rancho Nuevo; Río Cancita, 14 km. E of Apatzingán (5); Río Marquez, 10 km. S of Lombardia (2); Río Marquez, 13 km. SE of Nueva Italia (3); San Salvador (2); Santa Ana; Tepalcatepec; Volcán Jorullo (3); Zicuiran (2); Ziracuaretiro.

Although individuals of this species have been collected at elevations slightly exceeding 1200 meters on Volcán Jorullo and at 1100 meters at Ziracuaretiro on the southern slopes of the Cordillera Volcánica, for the most part these lizards are found at elevations of less than 800 meters, where they inhabit the open arid scrub forest of the Tepalcatepec Valley, a region to which this species is endemic (Duellman, 1958b:49). These small lizards usually are found on the trunks and main branches of the small trees in the scrub forest; in this habitat they are associated with *Sceloporus horridus oligoporus*, a much larger species.

Males have a pale orange spot on the throat and a pale blue belly; females have immaculate venters.

A specimen from Guayabo on the northern slopes of the Sierra de Coalcomán was referred to *Urosaurus irregularis* (Fischer) by Peters (1954:15). I have studied this specimen (BMNH 1914.1.28.110), a female having a snout-vent length of 46 mm., and agree with Peters that it closely resembles Fischer's description and figure (1882: pl. 17, fig. 1). This specimen and those seen of *Urosaurus gadowi* all have pavementlike enlarged dorsal scales that are complete across the vertical line. In *U. gadowi* the enlarged dorsals usually are in four to six irregular rows; in the specimen from Guayabo the dorsals are in two rows. Although none of the other specimens of *U. gadowi* examined has only two rows of enlarged dorsals, I prefer to consider the specimen from Guayabo as an aberrant individual of that species, rather than *U. irregularis*. Guayabo is in the known range of *U. gadowi*. *Urosaurus irregularis* is known only from the type specimen in the Bremen Museum; the type locality, according to Fischer (1882:232), is "Aus dem Hochlande von Mexico." If an examination of the type specimen of *U. irregularis* shows it to be identical with *U. gadowi*, then *U. irregularis* would be the name for the lizards here referred to *U. gadowi*.

MABUYA BRACHYPODA TAYLOR

Mabuya brachypoda Taylor, Univ. Kansas Sci. Bull., 38 (1):308, December 20, 1956.—Four kilometers east-southeast of Los Angeles de Tilarán, Guanacaste, Costa Rica.

El Sabino (42); La Placita; Playa Azul; Tzitzio (3).

Previously this species has been reported from La Placita as *Mabuya mabouya alliacea* by Peters (1954:15). Webb (1958:1311) provided evidence that Mexican specimens were conspecific with *Mabuya brachypoda*, as described from Costa Rica by Taylor (1956:308). The large series in the Taylor collection studied by Webb and listed by him as being from Uruapan actually is part of a series collected by Hobart M. Smith at El Sabino at an elevation of 1050 meters, 30 kilometers south of Uruapan.

This species probably ranges throughout the coastal region of the state; individuals from La Placita and Playa Azul were taken in dense scrub forest near sea level.

SCINCELLA ASSATA TAYLORI (OLIVER)

Leiolopisma assatum taylori Oliver, Occ. Pap. Mus. Zool. Univ. Michigan, 360:12, November 20, 1937.—Santiago, Colima, México.

Scincella assata taylori, Mittleman, Herpetologica, 6:20, June 5, 1950.

Twenty-one km. S of Arteaga; Ostula.

The specimen from Ostula was obtained in semi-deciduous broad-leaf forest at an elevation of 120 meters; that from 21 kilometers south of Arteaga was taken in oak forest at an elevation of 830 meters. Both localities are on the coastal slopes of the Sierra de Coalcomán. Probably the species inhabits the heavy forests on the lower slopes of these mountains. The specimen from south of Arteaga (UMMZ 119117) in life had a tan dorsum and a bright orange-pink tail.

EUMECES ALTAMIRANI DUGÈS

Eumeces altamirani Dugès, La Naturaleza, ser. 2, 1:485, 1891.—Apatzingán, Michoacán, México.

Twelve km. E of Apatzingán; El Sabino (4).

One specimen of this rare species was found beneath a rock in the open scrub forest 12 kilometers east of Apatzingán on July 3, 1955. Another skink, presumably of this species, was seen at Capirio. The specimen from east of Apatzingán is a male having a snout-vent length of 97 mm. and an incomplete tail. In most respects it compares favorably with accounts of the species given by Taylor (1936b:55 and 1936c:102). The frontal is divided by a transverse suture; the enlarged dorsal scales are arranged in 11 pairs anteriorly, followed by 48 unpaired enlarged scales. The head and middorsal area are brown; there is a pale tan stripe on the edges of the vertebral and paravertebral rows, bordered by a dark brown stripe on the paravertebral row, which, in turn, is bordered by a pale tan stripe on the lateral edge of the paravertebral scale row and the median edge of the adjacent scale row. The stripes extend from the neck to the base of the tail. The flanks are mottled with brown and cream-color; the labials are cream-color barred by brown; the venter is a pale cream-color.

Dugès (1891:485) described *Eumeces altamirani* from "las regiones cálidas del Estado de Michoacán" and subsequently (1896:480) gave Apatzingán as a locality for the species. Presumably he had only one specimen. In 1935 Hobart M. Smith collected the species at El Sabino on the lower slopes of the Cordillera Volcánica bordering the Tepalcatepec Valley. All of the known specimens are from this valley and the adjacent slopes, an area to which the species apparently is endemic.

EUMECES COLIMENSIS TAYLOR

Eumeces colimensis Taylor, Publ. Field Mus. Nat. Hist., zool. ser., 20:77, May 15, 1935.—Colima, Colima, México.

Coalcomán; Salitre de Estopila.

The species was reported by Peters (1954:16); no additional material has been discovered. The species is known only from foothills and low mountains at elevations between 130 and 950 meters in Michoacán and Colima.

EUMECES COPEI TAYLOR

Eumeces copei Taylor, Proc. Biol. Soc. Washington, 46:133, June 5, 1933.—10 miles southeast of Asunción, México, México. Cerro Tancítaro (3); Zacapu.

This member of the *Eumeces brevirostris*-group has been found only in pine or pine-fir forests at elevations from 1800 to 2700 meters. It probably ranges throughout the high mountains of the state north of the Tepalcatepec Valley; its apparent absence in other parts of the Cordillera Volcánica, other than on Cerro Tancítaro, is surprising. The species has been taken near Asunción in the state of México and at Lagunas de Zempoala in Morelos.

In this species the lateral pale yellow stripe, which is bordered below by dark brown, extends to the groin and onto the base of the tail. The dorsolateral stripe is separated from the copper-colored middorsum by a narrow brown stripe.

EUMECES DUGESI THOMINOT

Eumeces Dugesii Thominot, Bull. Soc. Philom. Paris, ser. 7, 7:138, 1883.—Guanajuato. Type locality restricted to Guanajuato, Guanajuato, México, by Smith and Taylor (1950b:169).

Carapan (6); Cherán (5); Opopeo (2); 17 km. S of Pátzcuaro (3); San José de la Cumbre (2); Tancítaro (2); Tangancícuaro; Uruapan; Zacapu.

Individuals of this species frequently have been found beneath rocks and logs in pine-oak, pine, or fir forests from elevations of 1550 to 1850 meters. To judge from specimens available, *E. dugesi* probably is the most abundant and widespread species of skink in the state.

In this species the lateral yellow stripe is indistinct and is persistent only in the axilla; the dorsolateral stripes terminate anterior to the hind limbs and are not separated from the tan dorsum.

EUMECES INDUBITUS TAYLOR

Eumeces indubitus Taylor, Univ. Kansas Sci. Bull., 21:257, November 27, 1934.—Near Cuernavaca, Morelos, México.

Puerto Hondo.

The one specimen of this species from Michoacán was collected by Edward H. Taylor in pine forest at Puerto Hondo, near Zitácuaro, at an elevation of about 2750 meters (Taylor, 1935:466). The species is known from the high mountains of eastern Michoacán, western México, and northern Morelos.

EUMECES PARVULUS TAYLOR

Eumeces parvulus Taylor, Proc. Biol. Soc. Washington, 46:175, October 26, 1933.—Tepic, Nayarit, México.

El Ticuiz; La Placita; Pómaro (2); San Pedro Naranjestila (3).

Aside from the specimens reported by Peters (1954:17), one other specimen was obtained at El Ticuiz. It has 22 scale rows, 3 supraoculars in contact with the frontal, 2 postlabials, and a unicolored olive-tan dorsum. In life the anterior dorsolateral stripes were pale pinkish tan, the labials cream color, the throat white, and the tail pale blue. All specimens were found in semi-deciduous broad-leaf forest at elevations of less than 500 meters on the seaward slopes of the Sierra de Coalcomán.

AMEIVA UNDULATA SINISTRA SMITH AND LAUFE

Ameiva undulata sinistra Smith and Laufe, Univ. Kansas Sci. Bull., 31 (1):59, May 1, 1946.—Manzanillo, Colima, México.

Apatzingán (9); 19 km. S of Arteaga (3); Barranca de Bejuco (2); Coahuayana (6); Coalcomán (3); El Ticuiz (10); La Placita (2); Limoncito (3); Ostula (2); Playa Azul; Salitre de Estopila; San Juan de Lima (2); San Pedro Naranjestila (4).

Six males and six females from the Tepalcatepec Valley have more femoral pores than do 16 males and nine females from the coastal lowlands; the ranges and average number of femoral pores in the former are 40-50 (44.8) for males and 38-40 (38.6) for females; males from the coast have 34-44 (39.2), and females have 32-40 (36.2) femoral pores. In all specimens the number of lamellae beneath the fourth toe varies from 26 to 33 (29.7). In life juveniles have a pale olive-tan dorsum and a dorsolateral dark band, superimposed on which is a row of darker brown spots. The dorsolateral band is bordered below by a narrow cream-colored stripe. The tail is tan above and grayish white below; the belly is pale bluish white. Adult males are brilliantly colored in life. A male having a snout-vent length of 108 mm. had a rusty brown dorsum and bright blue bars on the flanks separated by dark brown interspaces. The side of the head was pale green, and the chin and throat were golden yellow. In some specimens the throat is orange. Juveniles and sub-adults have dark flecks on the brown or tan middorsal area, but these are absent in the largest males.

This species inhabits the heavily wooded areas in the lowlands to elevations of about 950 meters. In the Tepalcatepec Valley it has been found only in gallery forests along streams. In both the Tepalcatepec Valley and the coastal lowlands there is a noticeable absence of large adults in the dry season.

CNEMIDOPHORUS CALIDIPES DUELLMAN

Cnemidophorus calidipes Duellman, Occ. Pap. Mus. Zool. Univ. Michigan, 574:1, December 23, 1955.—Capirio, Michoacán, México.

Apatzingán (56); 12-20 km. S of Apatzingán (5); 19 km. E of Apatzingán (5); 25 km. S of Arteaga; Capirio (57); El Espinal (13); Jazmin (9); 11 km. S of Lombardia; Nueva Italia.

This small, distinctive species of the *sexlineatus*-group of *Cnemidophorus* was discovered in the Tepalcatepec Valley in 1955 (Duellman, 1955); subsequent field studies showed it to be widespread in the valley (Duellman, 1960c). One specimen (KU 29747) is from the relatively arid, low Pacific slope of the Sierra de Coalcomán, 25 kilometers south of Arteaga. All other specimens have been taken at elevations of 200 to 650 meters in the Tepalcatepec Valley, where the species characteristically inhabits the open scrub forests of the valley floor, especially the *Cercidium-Prosopis-Apoplanesia* associations, where there is a sparse growth of grasses. In this habitat it is most frequently seen in association with *Cnemidophorus costatus zweifeli* and *C. deppei infernalis.*

Aside from the characters given in Table 5, *Cnemidophorus calidipes* differs from other species of *Cnemidophorus* in Michoacán by possessing a complete (or nearly so) supraorbital semicircle-series of granules; in other species the granules seldom extend anteriorly beyond the posterior border of the frontal.

CNEMIDOPHORUS COMMUNIS COMMUNIS COPE

Cnemidophorus communis Cope, Proc. Amer. Philos. Soc., 17:95, 1877.—No type locality given; type locality restricted to Colima, Colima, México, by Zweifel (1959a:74).

Cnemidophorus communis communis, Zweifel, Bull. Amer. Mus. Nat. Hist., 117:74, April 27, 1959.

Aguililla (2); Apatzingán (6); 13 km. S of Arteaga (2); 19 km. S of Arteaga (3); Capirio (3); Coahuayana (3); Coalcomán (44); El Ticuiz; between El Ticuiz and Ojos de Agua de San Telmo; La Placita (6); Pómaro (2); Río Cachán; Salitre de Estopila; San Juan de Lima.

The specimens from Coalcomán and the coastal localities were referred to *Cnemidophorus sacki copei* by Peters (1954:18) and Duellman (1954b:12). Zweifel (1959a) referred these specimens to *Cnemidophorus communis communis* and pointed out the probable sympatry of *C. communis* and *C. costatus* (= *sacki* of Zweifel) in the Tepalcatepec Valley.

There is considerable geographic variation in the number of dorsal granules around the midbody. Sixteen specimens from the coastal regions of Michoacán have 129-146 (136.3) granules; nine from the Tepalcatepec Valley have 124-137 (128.3), and 44 from Coalcomán at an elevation of 950 meters in the Sierra de Coalcomán, intermediate geographically between the coast and the Tepalcatepec Valley, have 105-144 (119.7). The number of granules in specimens from the coast of Michoacán compares favorably with the range of 118-154 (137.8) for 34 specimens from Colima, Colima (Zweifel, 1959a:107). Aside from the characters given in Table 5, *C. communis communis* can be distinguished from other members of the *Cnemidophorus sexlineatus*-group (*calidipes*, *costatus*, and *scarlaris*) by its relatively small post-antebrachial scales.

TABLE 5.—COMPARISON OF THE TEN SPECIES AND SUBSPECIES OF CNEMIDOPHORUS IN MICHOACÁN (SCALE COUNTS ARE FOR SPECIMENS FROM MICHOACÁN ONLY)

Species	Dorsal granules	Femoral pores	Adult color pattern	Throat color	Maximum snout-vent length
calidipes	66-86 (75)	31-47 (39)	Light brown dorsum with vertical blue bars and spots	Pink	79 mm.
communis communis	105-146 (124)	38-52	Green dorsum with six rows of yellow spots	Pink	135 mm.
costatus occidentalis	97-102 (99)	37-43 (39)	Cross-bars anteriorly and pale spots posteriorly	Pink	126 mm.
costatus zweifeli	91-117 (106)	32-49 (41)	Lateral and dorsolateral rows of spots; paravertebrals fused with pale green middorsum	Pink with blue spot	132 mm.

deppei deppei	116-117 (116)	37-38 (37)	Green paravertebral and dorsolateral stripes; lateral stripe broken into row of bluish spots	Black	93 mm.
deppei infernalis	91-120 (101)	31-43 (36)	Green paravertebral and dorsolateral stripes; broad cream lateral stripe; reddish flanks	Black	84 mm.
lineatissimus exoristus	108-140 (122)	32-47 (39)	Paravertebral stripes fused with yellow middorsal stripe; vertical bars on flanks	Pink and black	98 mm.
lineatissimus lineatissimus	117-126 (121)	32-37 (35)	Eight distinct stripes plus partially fused vertebrals	Bluish-pink and black	96 mm.
lineatissimus lividus	126-164 (148)	32-48 (38)	Broad mid-dorsal stripe; paravertebrals distinct; blue lateral spots	Pink and black	106 mm.
scalaris	80-92 (86)	32-41 (35)	Six distinct cream stripes; tan spots in dark fields	Orange-pink	95 mm.

Although this is the largest species of *Cnemidophorus* in Michoacán (adult males attain a snout-vent length of 135 mm.), it is neither widespread nor abundant. On the coastal lowlands it occurs primarily with *Cnemidophorus lineatissimus lividus*. In the coastal lowlands there is little open scrub forest, a type of habitat that seems to be preferred by *C. communis communis*. In the Tepalcatepec Valley, *C. communis communis* occurs in the open scrub forest with the more abundant large species *C. costatus* (subspecies *zweifeli*). Only in the scrub forest in the Coalcomán Valley, where no other species of *Cnemidophorus* occurs, is *C. communis communis* abundant.

CNEMIDOPHORUS COSTATUS OCCIDENTALIS GADOW

Cnemidophorus communis occidentalis Gadow, Proc. Zool. Soc. London, 1906, 1:339, August 23, 1906.—Type locality restricted to Ixtlán, Nayarit, México, by Smith and Taylor (1950b:182).

Cnemidophorus costatus occidentalis, Zweifel, Copeia, No. 1:98; March 17, 1961.

Jiquilpan (4).

Only four specimens from the extreme northwestern part of the state are referable to this subspecies. These have 97 to 102 dorsal granules at midbody and lack the blue gular band or spot characteristic of the subspecies in the Tepalcatepec Valley. Probably *C. costatus occidentalis* ranges throughout the Chapala depression, but to the east it is replaced by *Cnemidophorus scalaris scalaris*.

CNEMIDOPHORUS COSTATUS ZWEIFELI DUELLMAN

Cnemidophorus sacki zweifeli Duellman, Univ. Kansas Publ. Mus. Nat. Hist., 10:589, May 2, 1960.—Capirio, Michoacán, México.

Apatzingán (107); Buenavista (3); Capirio (31); Charapendo (12); Chinapa (2); 19 km. S of Corralito (3); Jazmin (2); between La Playa and Volcán Jorullo (2); Limoncito (3); 14 km. S of Lombardia (11); Nueva Italia (15); Río Marquez, 10 km. S of Lombardia (2); Río Marquez, 13 km. SE of Nueva Italia; Tafetan (18); 14 km. E of Tepalcatepec (2); Tzitzio (11); 19 km. S of Tzitzio; Volcán Jorullo (5); Ziracuaretiro; Zirimícuaro.

These lizards were referred to *Cnemidophorus sacki copei* by Duellman (1954b:12 and 1955:6); Duellman (1960a) described the subspecies *zweifeli* and assigned it to *Cnemidophorus sacki*. Zweifel (1961:98) used the specific name C. *costatus* for the whiptails on the southwestern part of the Mexican Plateau (*C. c. occidentalis*). Since *occidentalis* and *zweifeli* are conspecific, the combination *C. costatus zweifeli* is used here for the population inhabiting the Tepalcatepec Valley.

This lizard is abundant in the Tepalcatepec Valley, where it lives in open and dense scrub forest, usually at elevations of less than 1000 meters. Throughout the valley it is found in association with *Cnemidophorus deppei infernalis*, and in the lower parts of the valley it also is associated with *Cnemidophorus calidipes*. Observations made in the dry season indicate that large adults are not active at that time.

On the coastal lowlands and in the valleys in the Sierra de Coalcomán *Cnemidophorus costatus zweifeli* is replaced by *C. communis communis*. To the east in the Balsas Basin *C. costatus zweifeli* intergrades with *C. costatus costatus*.

CNEMIDOPHORUS DEPPEI DEPPEI WIEGMANN

Cnemidophorus deppei Wiegmann, Herpetologia Mexicana, p. 29, 1834.—México. Type locality restricted to Tehuantepec, Oaxaca, México, by Smith and Taylor (1950b:179).

Cnemidophorus deppei deppei, Cope, Trans. Amer. Philos. Soc.,

17:31, 1892.

Salitre de Estopila; San Pedro Naranjestila.

This small species, which is extremely abundant on the coastal lowlands of Guerrero, seems to be rare on the coast of Michoacán, where it has been taken at elevations of 130 and 500 meters in open situations in otherwise forested areas. Duellman and Wellman (1960:25) discussed these specimens in relation to their subspecific assignment. They were referred to *Cnemidophorus deppei lineatissimus* by Peters (1954:18).

CNEMIDOPHORUS DEPPEI INFERNALIS DUELLMAN AND WELLMAN

Cnemidophorus deppei infernalis Duellman and Wellman, Misc. Publ. Mus. Zool. Univ. Michigan, 111:32, February 10, 1960.— Mexcala, Guerrero, México.

Acahuato; Apatzingán (227); Capirio (3); El Sabino; Jazmin; La Playa (6); Lombardia (6); Nueva Italia (4); Río Marquez, 10 km. S of Lombardia (6); Río Marquez, 13 km. SE of Nueva Italia (10); south of Tancítaro; Volcán Jorullo (3).

This is one of the most abundant and widespread lizards in the Tepalcatepec Valley. Throughout its range it is ecologically associated with *Cnemidophorus costatus zweifeli*, which ranges to elevations somewhat higher than the 1050 meters known for *C. deppei infernalis*. This small lizard reaches its greatest abundance in grassy areas on the floor of the Tepalcatepec Valley, where in the *Cercidium-Prosopis-Apoplanesia* associations it occurs with *Cnemidophorus calidipes*.

Duellman and Wellman (1960) discussed the variation and relationships of *Cnemidophorus deppei*, of which the subspecies *infernalis* is restricted to the Balsas-Tepalcatepec Basin.

CNEMIDOPHORUS LINEATISSIMUS EXORISTUS DUELLMAN AND WELLMAN

Cnemidophorus lineatissimus exoristus Duellman and Wellman, Misc. Publ. Mus. Zool. Univ. Michigan, 111:44, February 10, 1960.—Rancho Santa Ana, four kilometers northeast of San Salvador, Michoacán, México.

Thirteen to 25 km. S of Arteaga (18); Capirio (19); Limoncito (13); Santa Ana (22).

As in *Cnemidophorus calidipes*, the distribution of this subspecies seems to be restricted to the Tepalcatepec Valley, except in the vicinity of Arteaga, where it occurs on the southern slope of the Sierra de Coalcomán. As pointed out by Duellman and Wellman (1960:46), the specimens from south of Arteaga are like those from the Tepalcatepec Valley in scutellation and coloration, and not like *Cnemidophorus lineatissimus lividus* from the geographically closer coastal lowlands.

In the Tepalcatepec Valley *Cnemidophorus lineatissimus exoristus* inhabits gallery forests along the larger streams; in this habitat it is associated with *Ameiva undulata sinistra.* From the other species of *Cnemidophorus* in Michoacán, *C. lineatissimus exoristus* can be distinguished by the possession of seven longitudinal stripes in adults and by the characters of scutellation given in Table 5.

CNEMIDOPHORUS LINEATISSIMUS LINEATISSIMUS COPE

Cnemidophorus lineatissimus Cope, Proc. Amer. Philos. Soc., 17:94, 1877.—Colima and Guadalajara. Type locality restricted to Colima, Colima, México, by Smith and Taylor (1950b:179).

Cnemidophorus lineatissimus lineatissimus, Duellman and Wellman, Misc. Publ. Mus. Zool. Univ. Michigan, 111:41, February 10, 1960.

Boca de Apiza (4).

These specimens have 117 to 126 dorsal granules at midbody, a noticeably lower count than that for *Cnemidophorus lineatissimus lividus* on the coast of Michoacán, which has 126 to 164 (148). Apparently these specimens represent immature *C. lineatissimus lineatissimus;* the differences between these and *C. lineatissimus lividus* from nearby localities indicate that possibly the populations are distinct species and not subspecies, as suggested by Duellman and Wellman (1960:41).

CNEMIDOPHORUS LINEATISSIMUS LIVIDUS DUELLMAN AND WELL-MAN

Cnemidophorus lineatissimus lividus Duellman and Wellman, Misc. Publ. Mus. Zool. Univ. Michigan, 111:50, February 10, 1960.—Maruata, Michoacán, México.

Barranca de Bejuco (4); Boca de Apiza (2); Coahuayana (6); El Ticuiz (7); La Placita (11); Maruata (7); Motín del Oro; Ostula (5); Playa Azul (4); Playa Cuilala (2); Pómaro (2); Salitre de Estopila (2); San Pedro Naranjestila.

This is the most abundant and widespread species of *Cnemidophorus* on the coastal lowlands of Michoacán, where it ranges from sea level to elevations of about 500 meters. In this area it inhabits dense arid scrub forest and semi-deciduous broad-leafed forest. Both of these habitats are continuous, or nearly so, along the lowlands and foothills of the Sierra de Coalcomán. This in itself may explain the abundance of *Cnemidophorus lineatissimus* and the relative scarcity of *C. deppei* and *C. communis* in the coastal area, for *C. deppei* and *C. communis* usually inhabit more open arid scrub forest, as occurs in the Tepalcatepec Valley. Living in the dense scrub forest with *C. lineatissimus* is *Ameiva undulata sinistra.*

CNEMIDOPHORUS SCALARIS COPE

Cnemidophorus gularis scalaris Cope, Trans. Amer. Philos. Soc., 17:47, 1892.—Chihuahua, Chihuahua, México.

Cnemidophorus scalaris, Zweifel, Bull. American Mus. Nat. Hist., 117:72, 1959.

Araro (2); Jacona; Lago de Cuitzeo (42); Morelia; 21 km. N of Morelia (4).

Zweifel (1959a:72) assigned the small species of *Cnemidophorus* having a relatively low number of dorsal granules and inhabiting the southern part of the Mexican Plateau to *C. scalaris*, which he diagnosed as rarely exceeding 100 mm. in snout-vent length and always having an average of less than 100 dorsal granules at midbody and usually less than 90. Forty-two specimens from the south shore of Lago de Cuitzeo (UMMZ 119558) have 80-91 (85.8) dorsal granules. Four specimens from 21 kilometers north of Norelia (UIMNH 6952 and UMMZ 104743) have 89, 78, 92, and 84 granules; one from Morelia (UMMZ 104742) has 78; two from Araro (UMMZ 119522) have 80 and 87; one from Jacona (UIMNH 24703) has 88.

Since no large adult males are present in the series from Michoacán, an adequate comparison of coloration between these and populations on the northern part of the Mexican Plateau cannot be made. *Cnemidophorus scalaris* is a name applied to the lizards inhabiting the Mexican Plateau from Chihuahua south to Puebla by Zweifel (1959a:72). It is doubtful if all of the populations assigned to this subspecies belong there; possibly more than one species is involved, but the paucity of material prevents further analysis at this time.

HELODERMA HORRIDUM HORRIDUM (WIEGMANN)

Trachyderma horridum Wiegmann, Isis von Oken, 22:421, 1829.—México. Type locality restricted to Huajintlán, Guerrero, México, by Smith and Taylor (1950b:193).

Heloderma horridum horridum, Bogert and Martín del Campo, Bull. Amer. Mus. Nat. Hist., 109:20, April 16, 1956.

Apatzingán; Coalcomán; La Placita; Oropeo; Parácuaro.

This species is known from elevations of less than 1000 meters in the Tepalcatepec Valley, the Sierra de Coalcomán, and the coastal lowlands. Specimens from Coalcomán, La Placita, and Parácuaro came from areas of dense woods; those from Apatzingán and Oropeo might have come from patches of dense woods in the otherwise open scrub forest of the Tepalcatepec Valley.

GERRHONOTUS IMBRICATUS IMBRICATUS WIEGMANN

Gerrhonotus imbricatus Wiegmann, Isis von Oken, 21:379, 1828.—México. Type locality restricted to México, Distrito Federal, by Smith and Taylor (1950b:201).

Gerrhonotus imbricatus imbricatus, Dunn, Proc. Acad. Nat. Sci. Philadelphia, 88:475, October 20, 1936.

Acuaro de las Lleguas (9); Cerro Barolosa (4); Cerro Tancítaro (36); Dos Aguas (22); Paracho; Sierra Patamba; Tinguidín; Zacapu.

Specimens from the Sierra de Coalcomán are noticeably different from those inhabiting the mountains rising from the Mexican Plateau. Of 45 specimens from Cerro Tancítaro and adjacent areas on the Mexican Plateau and in the Cordillera

Volcánica, 15 have twelve longitudinal rows of dorsal scales and 30 have fourteen rows. Of seven specimens from the state of México, 5 have twelve rows and 2 have fourteen; of nine specimens from central Veracruz, 8 have twelve rows and one has fourteen; of six specimens from Hidalgo, 5 have twelve rows and one has sixteen; of two specimens from Guanajuata, one has fourteen and the other has sixteen rows. All of the 35 specimens from the Sierra de Coalcomán have sixteen rows. Furthermore, these specimens have the superciliary row extended anteriorly, so that the anterior superciliary is in broad contact with the loreal. Specimens from Cerro Tancítaro have a shorter superciliary row, so that the anterior superciliary is not in broad contact with the loreal. These characters were used by Tihen (1949:220) to distinguish *Gerrhonotus imbricatus ciliaris* from *G. imbricatus imbricatus*. According to Tihen, the subspecies *G. imbricatus ciliaris* ranges from Guanajuato and Hidalgo northward to Chihuahua and Coahuila, whereas the nominal subspecies occurs from Michoacán and Hidalgo southward to Oaxaca. Specimens from the Sierra de Autlán in Jalisco are like those from Cerro Tancítaro; consequently, there seems to be no connection between the populations of *G. imbricatus ciliaris* in the mountains of the northern part of the Mexican Plateau with the *ciliaris*-like individuals found in the Sierra de Coalcomán. The acquisition and study of additional material from throughout the range of the species is necessary to clarify the picture of geographic variation. Until then, I prefer to consider all of the specimens from Michoacán as *Gerrhonotus imbricatus imbricatus*.

The largest specimen is a male having a snout-vent length of 136 mm. Two juveniles collected in July 24, 1960, have snout-vent lengths of 36 and 42 mm. A specimen having a snout-vent length of 127 mm. and a tail length of 145 mm. was regurgitated by a *Crotalus pusillus*, which had a body length of 550 mm.

Gerrhonotus imbricatus imbricatus is an inhabitant of coniferous forests. In the Cordillera Volcánica it occurs from 1500 to 3500 meters at the top of Cerro Tancítaro. In the Sierra de Coalcomán it occurs from 2100 to 2700 meters. On July 4, 1955, a pair was found in copulation beneath a pine log at 2700 meters on Cerro Barolosa. The male was lying on top of the female and was holding her head firmly in his jaws; the male's tail was curled under the female's tail, so that the cloacae were in contact.

SERPENTES

TYPHLOPS BRAMINUS (DAUDIN)

Eryx braminus Daudin, Hist.... des reptiles, 7:279, 1803. — Vazagapatam, India.

Typhlops braminus, Cuvier, Règne animal, ed. 2, 2:73, 1829.

Apatzingán; Arteaga.

Both specimens known from Michoacán were collected by Gadow in 1908. Peters (1954:20) remarked that the specimen from Arteaga probably does not indicate a rapid spreading of the species, which most likely was introduced into México at the time that vessels were stopping at Acapulco from the Philippines (Taylor, 1940b:444), but instead may indicate that pack trains from Acapulco passed through the Sierra de Coalcomán. The occurrence of this snake along a long-used *camino* substantiates this belief.

LEPTOTYPHLOPS BRESSONI TAYLOR

Leptotyphlops bressoni Taylor, Copeia, No. 1:5, March 9, 1939.— Hacienda El Sabino, Michoacán, México.

El Sabino.

This species still is known definitely only from the type specimen collected on the lower slopes of the Cordillera Volcánica at the northern edge of the Tepalcatepec Valley. A specimen (now lost) reported from Aguililla by Cope (1887:63) possibly represents this species (see Smith and Taylor, 1945:21, and Peters, 1954:20).

LEPTOTYPHLOPS GADOWI DUELLMAN

Leptotyphlops gadowi Duellman, Copeia, No. 2:93, May 29, 1956.—Apatzingán, Michoacán, México.

Apatzingán.

No additional specimens of this species have been collected since the species was described by Duellman (1956b:93). Data given with the specimen by Gadow indicate that it came from his camp above Apatzingán at an elevation of about 800 meters. Although the exact position of this camp is unknown, the lower slopes of the Cordillera Volcánica above Apatzingán usually support arid scrub forest at elevations below 1000 meters. Therefore, this species probably is an inhabitant of the arid scrub forest.

LEPTOTYPHLOPS PHENOPS BAKEWELLI OLIVER

Leptotyphlops bakewelli Oliver, Occ. Pap. Mus. Zool. Univ. Michigan, 360:16, November 20, 1937.—Paso del Río, Colima, México.

Leptotyphlops phenops bakewelli, Smith, Proc. U. S. Natl. Mus., 93:445, October 29, 1943.

La Placita (4); La Salada; Ostula.

The five specimens from the coastal lowlands are from elevations of less than 150 meters; these were collected by Peters (1954:20); the specimen from La Salada is from an elevation of 580 meters in the Tepalcatepec Valley. Peters (*loc. cit.*) remarked that the rostral and the tip of the tail that were described as white by Oliver (1937:17) actually are sulphur-yellow in life.

LOXOCEMUS BICOLOR COPE

Loxocemus bicolor Cope, Proc. Acad. Nat. Sci. Philadelphia, 13:77, June 30, 1861.—La Unión, El Salvador.

Loxocemus sumichrasti Bocourt, Ann. Sci. Nat., ser. 6, 4:1,

1876.—Tehuantepec, Oaxaca, México.

Apatzingán (6); La Orilla; Lombardia.

As noted by Peters (1954:21), this species was not recorded from Michoacán by Smith and Taylor (1945:27), but Gadow (1930:30) collected a specimen at La Orilla in 1908. This specimen (BMNH 1914.1.28.124) is a male having 235 ventrals and 47 caudals, a dark brown dorsum, and cream-colored labials and venter. The anterior chin-shields are considerably longer than the scales bordering the chin-shields. In these characters this specimen agrees with the diagnosis of *Loxocemus bicolor* given by Taylor (1940c:447), who revived *Loxocemus sumichrasti* Bocourt. Of the six specimens from Apatzingán in the Tepalcatepec Valley, three males have 243 to 253 (246.6) ventrals and 44 to 45 (44.3) caudals; three females have 238 to 247 (244.0) ventrals and 42 to 44 (43.0) caudals. Certain characters of scutellation utilized by Taylor for separating *L. bicolor* and *L. sumichrasti* are inconsistent in this series. The chin-shields are longer than the adjacent scales, like those illustrated in *L. bicolor* by Taylor (*op. cit.*, fig. 1). The relative lengths of the prefrontal and internasal sutures are subequal, or the prefrontal suture is slightly longer. Thus, in these characters of scutellation these snakes are like *L. bicolor*, but in coloration they are like *L. sumichrasti*; the dorsal color in life was an iridescent dark bluish gray, and the belly was pale gray or bluish gray.

The supposed differences in scutellation between *L. bicolor* and *L. sumichrasti* have been questioned by Woodbury and Woodbury (1944:360); these authors treated *L. sumichrasti* as a subspecies of *L. bicolor*. As pointed out by Zweifel (1959b:5), such an arrangement is not tenable, for, although individuals with each kind of color pattern have not been collected together at any one locality, the over-all geographic picture is one of sympatric distribution. Only snakes having the coloration of *L. sumichrasti* have been collected in the Balsas-Tepalcatepec Basin. I agree with Zweifel (*loc. cit.*) that on the basis of morphological similarities and sympatric distribution, *L. bicolor* and *L. sumichrasti* seem to be dimorphic phases of the same species, showing no more striking differences in coloration than *Lampropeltis getulus californiae*, a now classical example of pattern dimorphism in snakes.

In Michoacán, as in other parts of its range, *Loxocemus bicolor* inhabits arid scrub forest environments at low elevations.

BOA CONSTRICTOR IMPERATOR DAUDIN

Boa imperator Daudin, Hist. nat.... des reptiles, 5:150, 1803.—México. Type locality restricted to Córdoba, Veracruz, México, by Smith and Taylor (1950a:347).

Boa constrictor imperator, Forcart, Herpetologica, 7:199, December 31, 1951.

Apatzingán (4); Coalcomán; El Sabino (2); La Placita; La Playa (2); Lombardia; Nueva Italia (2); Río Cachán; Río Marquez, 13 km. SE of Nueva Italia; Río Nexpa; Volcán Jorullo.

These specimens have come from a variety of habitats from elevations of less than 1,000 meters. The species seems to be equally abundant in the broad-leafed semi-deciduous forests of the coastal foothills and in the arid Tepalcatepec Valley. In the latter area most of the specimens were collected at night.

CONIOPHANES FISSIDENS DISPERSUS SMITH

Coniophanes fissidens dispersus Smith, Proc. U. S. Natl. Mus., 91:106, November 13, 1941.—El Limoncito, Guerrero, México.

Arteaga.

Further collecting in southern Michoacán has failed to add additional material of this species, which is known in the state from the one specimen collected by Gadow in 1908. The species possibly ranges throughout the coastal foothills of the Sierra de Coalcomán. Peters (1954:21) described the specimen from Arteaga.

CONIOPHANES LATERITIUS LATERITIUS COPE

Coniophanes lateritius Cope, Proc. Acad. Nat. Sci. Philadelphia, 13:524, March 31, 1862.—Guadalajara, Jalisco, México.

Coniophanes lateritius lateritius, Smith and Grant, Herpetologica, 14:20, April 25, 1958.

Nineteen km. S of Arteaga.

The one specimen available from Michoacán of this apparently rare species was discussed by Wellman (1959:127), who pointed out that although the specimen was geographically intermediate between the subspecies *C. l. lateritius* (Jalisco and Nayarit) and *C. l. melanocephalus* (Morelos and Puebla), the specimen (UMMZ 118954) was like *C. l. lateritius* in scutellation and in color pattern differed from other known specimens of the species in having had in life a pale orange, instead of a brick-red, dorsum. Additional specimens from the Sierra de Coalcomán will be required in order to determine whether this specimen is a representative of an orange-colored population or merely is aberrant in coloration.

The present specimen is from an elevation of 900 meters in oak forest on the southern slopes of the Sierra de Coalcomán; other locality records for the species indicate that it inhabits broad-leafed forest in foothills from Nayarit to Puebla.

CONOPHIS VITTATUS VITTATUS PETERS

Conophis vittatus Peters, Monats. Akad. Wiss. Berlin, p. 519, 1860.—No type locality given. Type locality restricted to Laguna Coyuca, Guerrero, México, by Smith and Taylor (1950a:331).

Conophis vittatus vittatus, Smith, Jour. Washington Acad. Sci., 31:119, March 17, 1941.

Arteaga; Coalcomán (4); La Playa; 19 km. S of Tzitzio.

All specimens of this terrestrial snake have been collected in areas of scrub forest between 800 and 1100 meters above sea level. Since the species is known from the coastal regions of Guerrero and Colima, its absence from the cost of Michoacán is unexplainable; probably the lack of specimens from these areas is due solely to inadequate collecting.

CONOPSIS BISERIALIS TAYLOR AND SMITH

Conopsis biserialis Taylor and Smith, Univ. Kansas Sci. Bull., 28 (2):333, November 12, 1942.—Ten miles west of Villa Victoria, México, México.

Capácuaro (5); Cerro San Andrés; Cherán; Ciudad Hidalgo; Macho de Agua (4): Pátzcuaro (8); Tancítaro (24); Uruapan (9); 24 km. SE of Zitácuaro (14).

This species is abundant in the coniferous forests at elevations from 1550 to 2800 meters throughout the Cordillera Volcánica; apparently it does not occur in the Sierra de Coalcomán.

On August 1, 1956, a copulating pair was found beneath a rock at Capácuaro.

One of the best characters to distinguish this species from *Toluca lineata*, which occurs with *Conopsis* throughout its range in Michoacán, is the presence of large, black ventral blotches in *Conopsis biserialis*, as contrasted with the two rows of small black spots in *Toluca lineata*.

CONOPSIS NASUS GÜNTHER

Conopsis nasus Günther, Catalogue... snakes... British Museum, p. 6, 1858.—California (in error). Type locality restricted to Guanajuato, Guanajuato, México, by Smith and Taylor (1950a:330).

Carapan (2); Erongaricuaro; Maravatio (3); Morelia (2); Nahuatzen; Pátzcuaro (7); Tacícuaro (8); Tancítaro.

This species has been collected in oak, pine-oak, and fir forests at elevations of 1900 to 2450 meters on the mountains rising from the Mexican Plateau. It does not seem to be so abundant as *Conopsis biserialis*. Sufficient ecological data to determine differences in habitat between the two species have not been compiled.

DIADOPHIS DUGESI VILLADA

Diadophis punctatus dougesii Villada, La Naturaleza, 3:226, 1875.—Potreros de Balbuena, Distrito Federal, México.

Diadophis dugesii, Blanchard, Bull. Chicago Acad. Sci., 7:51, December 30, 1942.

Morelia (2); Pátzcuaro; Quiroga.

Apparently this snake is uncommon in Michoacán. It has been found only at elevations of 1900 to 2200 meters in pine and pine-oak forests on the mountains rising from the Mexican Plateau.

DRYADOPHIS MELANOLOMUS STUARTI SMITH

Dryadophis melanolomus stuarti Smith, Proc. U. S. Natl. Mus.,

93:418, October 29, 1943.—Acapulco, Guerrero, México.

Coahuayana; El Ticuiz; La Placita (3); Punto San Juan de Lima; Punto San Telmo.

The few specimens indicate that in Michoacán, as elsewhere on the Pacific coast of México, this species is restricted to forested regions on the coastal plain. It does not occur in the Tepalcatepec Valley.

The coloration, in life, of a juvenile (UMMZ 114604) is as follows: The dorsum is uniform pale grayish tan on posterior one-third of body and on tail; anteriorly there are pale grayish tan middorsal blotches separated by grayish white interspaces, which are about one-half the length of the blotches. Posteriorly the blotches are less distinct, fading into the uniform grayish tan ground color of the posterior part of the body. The blotches extend laterally onto the fourth and fifth scale rows. Large squarish lateral intercalary blotches of darker brown interconnect with the dorsal blotches. The top of the head is pale olive-brown; a dark brown postorbital stripe extends from the eye to the posterior edge of the last upper labial. The labials, chin, and ventrals 1-30 are creamy white, changing to a dusty cream-color posteriorly; the chin and ventrals 1-30 are heavily spotted with dark brown. The iris is a cream-color above and chocolate brown below; the tongue is blue.

DRYMARCHON CORAIS RUBIDUS SMITH

Drymarchon corais rubidus Smith, Jour. Washington Acad. Sci., 31:474, November 11, 1941.—Rosario, Sinaloa, México.

Apatzingán (5); Arroyo El Salto; Arteaga; Capirio; El Sabino (7); La Palma; La Placita; Ostula; San Juan de Lima.

Not all of the specimens from Michoacán are typical in color pattern of this subspecies, as defined by Smith (1941a:475). All specimens from the Tepalcatepec Valley are uniformly black above; they have reddish or cream-colored chins and the anterior two-thirds of the belly salmon-pink or reddish buff. Individuals from the Sierra de Coalcomán (Arteaga and Arroyo El Salto) are like those from the Tepalcatepec Valley. Three specimens from the coastal lowlands differ noticeably in color pattern:

UMMZ 104504, adult male (Ostula).—Pale brown above flecked with black anteriorly; at midbody, flecks form narrow transverse bands that become progressively wider posteriorly, until on tail no brown pigment evident, all ventrals reddish buff, except last eight, which are black.

UMMZ 104602, adult female (La Placita).—Black above, reddish cross-bands and flecks on all of body; dorsal and ventral surfaces of tail black; chin cream-color and entire belly reddish buff.

UMMZ 114626, adult male (San Juan de Lima).—Black above; dull rust-colored cross-bands on anterior half of body; chin white; belly rust-colored on anterior two-thirds of body and black posteriorly.

One specimen from La Palma on the Mexican Plateau (KU 29275) has the top of the head an olive-color, the entire dorsum black, the chin and ventrals 1-42 a cream-color, remainder of venter black, and all of the labials heavily barred with black. A juvenile from Capirio in the Tepalcatepec Valley (UMMZ 114627) is black above and has pale olive-colored flecks on the anterior one-third of the body; the top of the head is dark olive-brown, and the sides of the head are somewhat paler. Anteriorly the belly is a cream-color; posteriorly it is black.

The specimens from the Tepalcatepec Valley are typical of *Drymarchon corais rubidus*. Those from the coastal lowlands differ in having large areas of brown or red pigment on the dorsum, a condition not mentioned by Smith in his description of the subspecies. The specimen from La Palma, like many others from various localities on the Mexican Plateau, resembles in certain characters *D. corais orizabensis* (Smith, *op. cit.*: 477). Our knowledge of the geographical variation in coloration in this species is incomplete; many populations have been assigned to subspecific rank without justification.

In Michoacán this species is found from sea level to 1350 meters in the Sierra de Coalcomán and to 1300 meters at La Palma on Lago de Chapala. It has been collected in scrub forest, semi-deciduous broad-leafed forest, and oak forest.

DRYMOBIUS MARGARITIFERUS FISTULOSUS SMITH

Drymobius margaritiferus fistulosus Smith, Proc. U. S. Natl. Mus., 92:383, November 5, 1942.—Miramar, Nayarit, México.

Apatzingán (3); Coahuayana; Coalcomán (3); El Sabino (3); El Ticuiz; 12 km. S of Tzitzio.

This snake is abundant in the lowlands of the state; the few specimens listed above are indicative not of the rarity, but rather of the speed and agility, of this diurnal snake. It most frequently is found near water, where there is a dense growth of vegetation. One individual was observed in a large pool inhabited by several small *Rana pipiens*, and another was seen along the bank of a hyacinth-choked river channel. A third individual was captured while it was in pursuit of a *Cnemidophorus*.

This species has been collected on the coastal lowlands and seaward foothills of the Sierra de Coalcomán and in the Tepalcatepec Valley to elevations of 1150 meters.

ELAPHE TRIASPIS INTERMEDIA (BOETTGER)

Pityophis intermedius Boettger, Ber. Offen. Vereins. Naturk., 22:148, 1883.—México. Type locality restricted to Hacienda El Sabino, Michoacán, México, by Dowling (1960:74).

Elaphe triaspis intermedia, Mertens and Dowling, Senckenbergiana, 33:201, November 15, 1952.

Twenty-four km. E of Apatzingán; Chupio; El Sabino (4); 11 km. E of Emiliano Zapata.

Dowling (1960) has shown that specimens from the Balsas-Tepalcatepec Basin have fewer ventrals and caudals than those from the Sierra del Sur or the coast.

All specimens from Michoacán were collected in open forest, either scrub or oak forest. They were found in drier situations than those described for the species in southern Tamaulipas by Martin (1958:69). In Michoacán *Elaphe triaspis intermedia* is known from the Tepalcatepec Valley, the lower slopes of the Cordillera Volcánica, and the western edge of the Mexican Plateau at an elevation of 1350 meters. It probably occurs in the lower parts of the Sierra de Coalcomán and along the Pacific coast, for it is known from the coastal lowlands of Guerrero and Colima. In August, 1951, I saw a snake that probably was this species in Barranca de Bejuco.

ENULIUS UNICOLOR (FISCHER)

Geophis unicolor Fischer, Abh. Nat. Ver. Bremen, 7:227, 1882.— México. Type locality restricted to Chilpancingo, Guerrero, México, by Smith and Taylor (1950a:331).

Enulius unicolor, Taylor and Smith, Univ. Kansas Sci. Bull., 25:247, July 10, 1939.

Between Ario de Rosales and La Playa; Coalcomán; Jungapeo (4); between Zitácuaro and Tuxpan.

This small snake has been collected from beneath rocks in brushy areas and broad-leafed forest between 900 and 1800 meters; it has not been found in coniferous forest. The limited ecological data suggest that the species inhabits the transition zone between the tropical scrub forest and the temperate hardwood forest.

All of the specimens have 17 rows of scales; four males have 169-178 (174.2) ventrals and 102-111 (106.8) caudals; two females have 192 and 195 ventrals and 96 and 87 caudals. Three individuals have one postocular on one side and two on the other; in the other specimens there are two postoculars on each side. The largest male has a body length of 232 mm. and a tail length of 130 mm.; the largest female has a body length of 274 mm. and a tail length of 119 mm.

GEAGRAS REDIMITUS COPE

Geagras redimitus Cope, Jour. Acad. Nat. Sci. Philadelphia, ser. 2, 8:141, 1876.

San Juan de Lima (2).

Previously this species was known definitely only from the Plains of Tehuantepec, Oaxaca. *Sphenocalamus lineolatus* was described by Fischer (1883:5) from Mazatlán; this name has been placed in the synonymy of *Geagras redimitus* Cope. Although Fischer gave the type locality only as "Mazatlán" and did not designate the state, it is probable that the type originated from Mazatlán, Sinaloa. The present specimens are from a locality almost midway between Tehuantepec and Mazatlán and support the possibility that *Geagras* ranges along the Pacific coast of México from Oaxaca to Sinaloa.

The two specimens from Michoacán (UMMZ 114446-7), both males, have 118 and 122 ventrals, 31 and 33 caudals, body lengths of 108 and 81 mm., and tail lengths of 20 and 15 mm. Both have 1-1 preoculars, 1-1 postoculars, 1-2 temporals,

6-6 upper labials, and 5-5 lower labials. In life, the dorsum was pale tan; the top of the head and the middorsal and lateral stripes were dark brown; the belly was white. The occipital spots were pale pinkish tan. Both specimens were found beneath rocks in tropical semi-deciduous forest at an elevation of 15 meters on the coastal plain.

GEOPHIS DUGESI BOCOURT

Geophis dugesii Bocourt, Miss. Scientifique au Mexique et dans l'Amerique Centrale, Rept., livr. 9:573, 1883.—Tangancícuaro, Michoacán, México.

Carapan; Tangancícuaro; Zacapu.

Aside from the three specimens listed above, there are two (SU 4407-8) bearing the data "Michoacán." Bocourt (1883:574) stated that the type specimen from Tangancícuaro had six or seven pale cross-bands on the anterior part of the body. An illustration, presumably of the same specimen, by Dugès (1884:Pl. 9) shows five distinct and one indistinct cross-bands. Of the four specimens that I have examined, none has more than three pale cross-bands, and one has only one indistinct cross-band. Two females have 154 and 158 ventrals and 38 and 37 caudals; two males have 150 and 151 ventrals and 43 and 42 caudals.

This species is known only from elevations between 1750 and 2050 meters on the southwestern edge of the Mexican Plateau in the state of Michoacán.

GEOPHIS INCOMPTUS DUELLMAN

Geophis incomptus Duellman, Occ. Pap. Mus. Zool. Univ. Michigan, 605:3, May 29, 1959.—Dos Aguas, Michoacán, México.

Dos Aguas (15).

This species, which seems to be related to *Geophis maculiferus*, is known only from the pine-oak forest in the vicinity of Dos Aguas (elevation 2100 meters) in the Sierra de Coalcomán. Aside from the five specimens comprising the type series, there are ten other specimens in the Museum of Zoology at the University of Michigan collected by Floyd L. Downs in July, 1960. Data from these specimens and those comprising the type series show that in this sample seven males have 146-153 (149.3) ventrals and 35-37 (36.0) caudals; eight females have 150-154 (152.4) ventrals and 29-34 (32.5) caudals. The largest specimen is a female with a body length of 344 mm. and a tail length of 53 mm.

GEOPHIS MACULIFERUS TAYLOR

Geophis maculiferus Taylor, Univ. Kansas Sci. Bull., 27:119, December 30, 1941.—Near Cicio [*sic*] = Tzitzio, Michoacán, México.

Tzitzio.

The type and only known specimen of *Geophis maculiferus* (UIMNH 25078) is a female having 140 ventrals and 30 caudals, dorsal scales in 15 rows, one postocular, and an anterior temporal. Only one other species in México has dorsal scales in 15 rows and has an anterior temporal; that species is *G. incomptus*, which differs from *G. maculiferus* in having six or seven lower labials, instead of five, and in having the edges of the ventrals dark, instead of a uniformly cream-colored belly.

The locality from which the specimen was obtained lies at an elevation of 1630 meters on the southern slope of the Cordillera Volcánica. At that elevation there is an interdigitation of arid tropical scrub forest and pine-oak forest; probably *Geophis maculiferus* inhabits the pine-oak forest.

GEOPHIS NIGROCINCTUS DUELLMAN

Geophis nigrocinctus Duellman, Occ. Pap. Mus. Zool. Univ. Michigan, 605:1, May 29, 1959.—Dos Aguas, Michoacán, México.

Dos Aguas (3).

The three specimens comprising the type series of the species were found beneath logs and in a stump in pine-oak forest at an elevation of 2100 meters. A discussion of the variation in these specimens and of probable relationships of the species was given by Duellman (1959). Floyd Downs spent several days at Dos Aguas in July, 1960; although he found ten specimens of *Geophis incomptus*, no further specimens of *G. nigrocinctus* were obtained.

GEOPHIS PETERSI BOULENGER

Geophis petersii Boulenger, Catalogue Snakes... British Museum, 2:321, September 23, 1894.—Mexico City. Type locality restricted to Pátzcuaro, Michoacán, México, by Smith and Taylor (1950a:335).

Cherán; Coalcomán; Morelia; Pátzcuaro (6).

This seems to be the most widespread species of *Geophis* in Michoacán. It has been found at elevations between 950 and 2350 meters, chiefly in pine or pine-oak forest. Boulenger (1894:321) described *Geophis petersi* from a specimen stated to be from Mexico City, a locality which probably is in error. The only localities from which the species is definitely known are those listed in this account.

Three males and five females from the Mexican Plateau and the Cordillera Volcánica have respectively 140-144 (141.7) and 143-151 (146.0) ventrals and 39-41 (40.0) and 29-35 (33.2) caudals. All have dorsal scales in 15 rows, 1 postocular, no anterior temporal, and a relatively small triangular supraocular. The specimen from Coalcomán (UMMZ 104698) was referred to *Geophis nasalis* by Peters (1954:22). This specimen is abnormal in several characters; in five places there is a fusion and separation of the vertebral and paravertebral scale rows, producing a change from 17 to 15 rows of dorsal scales. Fusion of the three rows takes place at the level of the 8th, 41st, 47th, 54th, and 65th ventrals. Furthermore, there is a small secondary postocular on each side of the head. In other characters the specimen is like *G. petersi*; the resemblances to that species are greater than to *G. nasalis*, which has been recorded from Guatemala and southern Chiapas.

GEOPHIS TARASCAE HARTWEG

Geophis tarascae Hartweg, Occ. Pap. Mus. Zool. Univ. Michigan, 601:1, May 4, 1959.—Uruapan, Michoacán, México.

Uruapan (3).

A female of this species was collected in the Parque Nacional at the north edge of Uruapan in 1899, and a male was taken there in 1947; these specimens were used by Hartweg in his description of the species. Floyd L. Downs obtained another specimen in the Parque Nacional on July 19, 1960. It has 164 ventrals and 46 caudals; in life, the ground color of the neck was brown with a purplish tint; the dorsal markings were black; the chin was a cream-color, and the belly was white. This specimen is distinguished from those of all other species of *Geophis* in Michoacán in that it has dark irregular cross-bars on the dorsum and a row of dark spots on the venter.

HYPSIGLENA TORQUATA OCHRORHYNCHA COPE

Hypsiglena ochrorhyncha Cope, Proc. Acad. Nat. Sci. Philadelphia, 12:246, November 15, 1860.—Cape San Lucas, Baja California, México.

Hypsiglena torquata ochrorhyncha, Bogert and Oliver, Bull. Amer. Mus. Nat. Hist., 83:378, March 30, 1945.

Tupátaro.

The systematic status of the geographic variants of *Hypsiglena* in México and southwestern United States has been commented on by several authors. Tanner (1944) considered *H. torquata* and *H. ochrorhyncha* to be distinct species; Bogert and Oliver (1945:379) and Duellman (1957b:238) presented evidence indicating that *H. torquata* and *H. ochrorhyncha* intergrade in Sinaloa and southern Sonora. In *Hypsiglena* the scutellation, including the numbers of labials, dorsals, ventrals, and caudals, seem to vary in a clinal manner. Nevertheless, these snakes can be divided into two distinct populations on the basis of the nuchal color pattern, consisting of an *ochrorhyncha*-type (a broad dark nape-band, the lateral edges of which extend anteriorly and fuse with a postorbital stripe, and a narrow nape stripe extending from the posteromedian edges of the parietals to the dark nape band) and a *torquata*-type (a somewhat narrower dark nape-band bordered anteriorly by a pale nuchal area, and no dark nape stripe). Snakes having the *ochrorhyncha*-type of nuchal pattern are found on the Mexican Plateau from Michoacán northward into the desert regions of Sonora and the southwestern United States. Snakes having the *torquata*-type of pattern are found on the coastal lowlands and adjacent slopes of the Sierra Madre Occidental from southern Sinaloa to Colima and thence inland in the Balsas-Tepalcatepec Basin to Morelos and Guerrero. An exception is *Hypsiglena torquata dunklei* from Forlón and San Fernando, Tamaulipas; it has the *torquata*-type of nuchal pattern. The distributional picture is somewhat complicated because some individuals having the *torquata*-type of nuchal pattern also have a faint nape stripe. If these are taken as exceptions, the general picture of distribution in México is *H. t. torquata* on the Pacific lowlands from Sinaloa southward to the Balsas Basin and *H. t. ochrorhyncha* on the Mexican Plateau.

Smith (1943:433) resurrected *Hypsiglena jani* Dugès for the snakes of the *ochrorhyncha*-type on the southern part of the Mexican Plateau. He stated that the southern specimens differed from northern ones in having a nuchal spot 9 or 10 scales in length, as compared with a spot 2 to 6 scales in length in northern specimens. A cursory examination of specimens from the areas between Arizona and Michoacán showed that there is a gradual increase in the size of the spot from north to south. If no other characters can be found to distinguish the populations, they should be considered as a single subspecies.

Hypsiglena affinis differs from *H. torquata* in possessing 19 instead of 21 rows of dorsal scales. Additional material is needed from the western slopes of Jalisco and the Barrancas in Zacatecas and Durango, before definite allocation of *affinis* can be made.

Bogert and Oliver (1945:379) discussed the status of certain named populations in Baja California and concluded that only one species occurs there, and that the species probably is conspecific with *H. torquata*. A careful review of the genus *Hypsiglena* might show that there is only one species.

The one specimen from Michoacán (USNM 46513) is from an elevation of about 2300 meters near the southern edge of the Mexican Plateau.

HYPSIGLENA TORQUATA TORQUATA (GÜNTHER)

Leptodeira torquata Günther, Ann. Mag. Nat. Hist., ser. 3, 5:170.—Laguna Island, Nicaragua (in error).

Hypsiglena torquata torquata, Taylor, Univ. Kansas Sci. Bull., 25:371, July 10, 1939.

Apatzingán; Capirio; Cofradía.

Specimens from the three mentioned localities have the dark nuchal spot bordered anteriorly by a pale blotch. In life the specimen from Capirio (UMMZ 114424) had rich reddish brown dorsal spots; the dorsal ground color was grayish white above and somewhat more gray laterally. The pale nuchal area was a cream-color, and the iris was grayish red.

All of the specimens were found in the arid scrub forest in the Tepalcatepec Valley at elevations between 200 and 350 meters.

IMANTODES GEMMISTRATUS GRACILLIMUS (GÜNTHER)

Dipsas gracillima Günther, Biol. Centrali-Americana, Rept., p. 177, July, 1895.—southern México. Type locality restricted to Acapulco, Guerrero, México, by Smith and Taylor (1950a:331).

Imantodes gemmistratus gracillimus, Zweifel, Amer. Mus. Novitates, 1961:12, September 16, 1959.

La Orilla.

The specimen from La Orilla was reported by Peters (1954:23) as *Imantodes gemmistratus oliveri*; Zweifel (1959c) showed that *I. g. oliveri* did not range west of Tehuantepec and that the snakes inhabiting the coastal lowlands of Guerrero, Michoacán, and Colima were assignable to the subspecies *gracillimus*. It may be assumed that this subspecies ranges throughout the coastal lowlands and foothills of the Sierra de Coalcomán.

IMANTODES GEMMISTRATUS LATISTRATUS (COPE)

Dipsas gemmistrata latistrata Cope, Bull. U. S. Natl. Mus., 32:68, 1887.—Southern Jalisco. Type locality restricted to Guadalajara, Jalisco, México, by Smith and Taylor (1950a:334).

Imantodes gemmistratus latistratus, Zweifel, Amer. Mus. Novitates, 1961:3, September 16, 1959.

El Sabino.

The one specimen from Michoacán was collected near the upper limits of the scrub forest on the slopes of the Cordillera Volcánica. Zweifel (1959c:10) stated that in certain aspects of coloration this specimen was like *I. gemmistratus gracillimus*, but in scutellation and other features of coloration it was like *I. g. latistratus*. There are too few specimens of this species to define the ranges of the various subspecies with any degree of accuracy, but from the limited number of specimens available, it seems that *I. gemmistratus gracillimus* occurs on the Pacific lowlands from Guerrero northward to Colima. Northward on the Pacific lowlands from Colima to Sinaloa and in the Balsas-Tepalcatepec Basin is found *I. gemmistratus latistratus*.

LAMPROPELTIS DOLIATA (LINNAEUS)

Coluber doliatus Linnaeus, Systema naturae, ed. 12, 1:379, 1766.—Charleston, South Carolina.

Lampropeltis doliata, Klauber, Copeia, No. 1:11, April 15, 1948.

Coalcomán (3); El Sabino; 24 km. W of Morelia; Río Nexpa; Uruapan.

The few specimens of this species from Michoacán show a wide range of variation; furthermore, the present systematic status of the subspecies of *Lampropeltis doliata* portrays an incongruous pattern of distribution. Specimens from the Sierra de Coalcomán have relatively narrow red bands that are not interrupted dorsally by extensions of the black rings; the scales in the red bands have black tips. The specimen from El Sabino (EHT-HMS 5253) and the one from the Río Nexpa on the coast (USNM 31491) have broader red bands; the scales in the red bands do not have black tips. A specimen from 24 kilometers west of Morelia (UIMNH 17782) and one from Uruapan (UMMZ 121508) have the red bands interrupted dorsally by extensions from the black rings.

Specimens from the Sierra de Coalcomán were referred to *L. doliata blanchardi* by Peters (1954:24), who noted that in some characters these snakes were like *L. d. nelsoni* and in others like *L. d. polyzona*. The individual from El Sabino was referred to *L. d. nelsoni* by Taylor (1940c:465); the one from 24 kilometers west of Morelia was referred to *L. d. arcifera* by Smith (1942c:198). If these assignments are correct, three subspecies of *Lampropeltis doliata* occur in Michoacán: *blanchardi* in the Sierra de Coalcomán, *nelsoni* on the coast and in the Tepalcatepec Valley, and *arcifera* on the Mexican Plateau and in the Cordillera Volcánica. Such a distribution is plausible, but the few specimens and our general lack of knowledge of the variation and relationships of the different populations do not permit a definite assignment at this time.

LAMPROPELTIS RUTHVENI BLANCHARD

Lampropeltis ruthveni Blanchard, Occ. Pap. Mus. Zool. Univ. Michigan, 81:8, April 28, 1920.—Pátzcuaro, Michoacán, México.

Morelia; Pátzcuaro; Tacícuaro.

At the present time this species is known definitely from only three localities on the Mexican Plateau in Michoacán. An incomplete skin from El Sabino (EHT-HMS 5438) was referred to this species by Taylor (1940c:465); the specimen cannot be found, so verification of the identification cannot be made at this time.

LEPTODEIRA LATIFASCIATA (GÜNTHER)

Hypsiglena latifasciata Günther, Biologia Centrali-Americana, Reptilia, p. 138, October, 1894.—Southern México. Type locality restricted to Huajintlán, Morelos, México, by Smith and Taylor (1950a:331).

Leptodeira latifasciata, Dunn, Proc. Natl. Acad. Sci., 22:696, December, 1936.

Apatzingán; El Sabino; La Playa; 32 km. E of Nueva Italia.

This nocturnal snake apparently ranges throughout the arid Balsas-Tepalcatepec Valley to elevations of about 1050 meters. It has been collected only in the arid scrub forest. Aside from the specimens listed by Duellman (1958a:93), there is one (UMMZ 120223) having eight body blotches, a body length of 510 mm. and a tail length of 103 mm.

LEPTODEIRA MACULATA (HALLOWELL)

Megalops maculatus Hallowell, Proc. Acad. Nat. Sci. Philadelphia, 13:488, March 31, 1862.—"Tahiti." Type locality restricted to Manzanillo, Colima, México, by Duellman (1958a:54).

Leptodeira maculata, Duellman, Bull. Amer. Mus. Nat. Hist., 114:53, February 24, 1958.

Aguililla (2); Apatzingán (24); Arteaga (2); Capirio (3); Charapendo (2); Coahuayana (3); Cofradía; Cuatro Caminos; La Placita (3); Lombardia (69); Nueva Italia (29); Pómaro; Río Marquez, 10 km. S of Lombardia (2); Salitre de Estopila; Tafetan (2); Volcán Jorullo.

This snake is abundant in the arid Tepalcatepec Valley; most of the specimens have been collected in arid scrub forest at elevations of less than 500 meters. With the onset of the rains in late June and early July, large numbers of these snakes can be found around temporary pools, where they feed on small frogs and toads. In the dry season few individuals were found, and all of those were beneath cover. Specimens from the coast have more body-blotches than do those from the Tepalcatepec Valley (Duellman, 1958a:56); otherwise the snakes show little variation.

LEPTODEIRA SPLENDIDA BRESSONI TAYLOR

Leptodeira bressoni Taylor, Univ. Kansas Sci. Bull., 25:321, July 10, 1939.—Hacienda El Sabino, Michoacán, México.

Leptodeira splendida bressoni, Duellman, Bull. Amer. Mus. Nat. Hist., 114:84, February 24, 1958.

Coalcomán (3); El Sabino (3); Uruapan (5).

The range of *Leptodeira splendida bressoni* apparently does not overlap that of *Leptodeira maculata*; the latter is restricted to the lower reaches of the arid scrub forest, whereas *L. s. bressoni* inhabits the upper limits of the arid scrub forest and the lower part of the pine-oak forest. Specimens have been collected between 950 and 1630 meters on the slopes of the Cordillera Volcánica and at 950 meters in the Sierra de Coalcomán. At Uruapan individuals were found beneath rocks along a stream and in a stone fence. *Leptodeira duellmani*, which was described from Coalcomán by Peters (1954:25), is an aberrant individual of *L. s. bressoni* (Duellman, 1958a:56).

LEPTOPHIS DIPLOTROPIS (GÜNTHER)

Ahaetulla diplotropis Günther, Ann. Mag. Nat. Hist., ser. 4, 9:25, 1872.—Tehuantepec, Oaxaca, México.

Leptophis diplotropis, Bocourt, Mission scientifique au Mexique et dans l'Amerique Centrale, Reptiles, livr. 15:835, 1897.

Between Aguililla and Dos Aguas; Arteaga; Coalcomán; El Diezmo; El Sabino (5); La Playa; Ocorla.

Most specimens of this species have been collected in tropical semi-deciduous forest at elevations of less than 1000 meters. In the Sierra de Coalcomán one was taken in pine-oak forest at an elevation of 1700 meters near Ocorla; another was found in broad-leafed forest between Aguililla and Dos Aguas at an elevation of 1600 meters. Most individuals have been seen in trees or bushes. The absence of broad-leafed forest in the Tepalcatepec Valley probably accounts for the absence of this snake in that area.

MANOLEPIS PUTNAMI (JAN)

Dromicus putnami Jan, Elenco sistematico degli Ofidi, p. 67, 1863.—San Blas, Nayarit, México.

Manolepsis putnami, Smith and Taylor, Bull. U. S. Natl. Mus.,

187:92, October 5, 1945.

La Placita (3); Maquili; Ostula.

In Michoacán the species has been found only in tropical semi-deciduous forest on the lower slopes of the Sierra de Coalcomán. From the observations made by Peters (1954:28), this snake is diurnal and feeds on teiid lizards.

MASTICOPHIS STRIOLATUS STRIOLATUS MERTENS

Coluber striolatus Mertens, Zoologica (Stuttgart), 32:190, 1934.—Substitute name for *Coluber lineatus* Bocourt, a secondary homonym of *Coluber lineatus* Linnaeus = *Lygophis lineatus*. Type locality restricted to Presidio de Mazatlán, Sinaloa, México, by Smith and Taylor (1950a:343).

Masticophis striolatus striolatus, Zweifel and Norris, Amer. Midl. Nat., 54:242, July, 1955.

Apatzingán (4); Arteaga; Coalcomán (3); El Sabino; Jiquilpan; La Palma; La Playa (3); Lombardia; Nueva Italia; Río Cachán; Santa Ana; Uruapan (2); Volcán Jorullo; Ziracuaretiro.

This large diurnal species inhabits open scrub forest and cultivated terrain from sea level to about 1650 meters. On the Mexican Plateau it is known from the area around Lago de Chapala, to which it possibly gained access through the valleys in the headwaters of the Tepalcatepec drainage. Specimens from southern Michoacán have been reported previously by Peters (1954:28) and Duellman (1954b:16) as *Masticophis flagellum lineatus*.

MASTICOPHIS TAENIATUS AUSTRALIS SMITH

Masticophis taeniatus australis Smith, Jour. Washington Acad. Sci., 31:390, September 11, 1941.—Guanajuato, Guanajuato, México.

Tacícuaro (2); Zamora.

This species reaches the southern limit of its distribution in the state of Michoacán. The limited ecological data available suggest that the species inhabits the open mesquite grassland of the Mexican Plateau.

OXYBELIS AENEUS AURATUS (BELL)

Dryinus auratus Bell, Zool. Jour., 2:324, 1825.—México. Type locality restricted to Tehuantepec, Oaxaca, México, by Smith and Taylor (1950a:340).

Oxybelis aeneus auratus, Bogert and Oliver, Bull. Amer. Mus. Nat. Hist., 83:381, March 30, 1945.

Coahuayana; El Sabino (4); between Las Tecatas and Las Higuertas; between Los Pozos and La Ciénega; Playa Azul; Pómaro (2); between Pómaro and Maruata (2); Punto San Telmo; Río

Tizupan.

On the basis of the number of specimens seen and collected on the seaward slopes of the Sierra de Coalcomán, this is a common snake there. Most specimens were collected in tropical semi-deciduous forest; others were collected in oak forest to an elevation of 1700 meters. Apparently *Oxybelis* does not inhabit the lower parts of the Tepalcatepec Valley; the only specimens from the inland area are four from El Sabino, which is situated at about 900 meters on the slopes of the Cordillera Volcánica. One individual was seen in gallery forest near Limoncito at an elevation of 730 meters on the northern slopes of the Sierra de Coalcomán.

PITUOPHIS DEPPEI DEPPEI (DUMÉRIL)

Elaphis deppei Duméril, Mem. Acad. Inst. France, 23:453, 1835.—México. Type locality restricted to San Juan Teotihuacán, México, México, by Smith and Taylor (1950a:334).

Pituophis deppei deppei, Stull, Occ. Pap. Mus. Zool. Univ. Michigan, 250:1, October 12, 1932.

Carapan (2); Morelia; Tacámbaro; Tacícuaro; Zacapu.

Duellman (1960b) showed that the widespread species *Pituophis deppei* was composite and that the "lined subspecies" actually represented another species, *Pituophis lineaticollis. Pituophis deppei* occurs only on the Mexican Plateau; in Michoacán it inhabits mesquite grassland and oak-bunch grass associations between 1900 and 2200 meters.

PITUOPHIS LINEATICOLLIS LINEATICOLLIS (COPE)

Arizona lineaticollis Cope, Proc. Acad. Nat. Sci. Philadelphia, 13:300, December 28, 1861.—Southern Mexican Plateau. Type locality restricted to 24 kilometers northwest of Oaxaca, Oaxaca, México, by Duellman (1960b:607).

Pituophis lineaticollis lineaticollis, Duellman, Univ. Kansas Publ. Mus. Nat. Hist., 10:607, May 2, 1960.

Acuaro de las Lleguas; Dos Aguas (3); Morelia; Tancítaro (5).

This species reaches the northern limits of its range in the Sierra de Coalcomán and on the Mexican Plateau in Michoacán. On the plateau it has been collected in mesquite grassland at elevations between 1500 and 2000 meters. In the Sierra de Coalcomán individuals were found in open pine-oak forest at 2100 meters elevation and in a meadow surrounded by pine-oak forest at 2300 meters.

PSEUDOFICIMIA FRONTALIS (COPE)

Toluca frontalis Cope, Proc. Acad. Nat. Sci. Philadelphia, 16:167, September 30, 1864.—Colima, Colima, México.

Pseudoficimia frontalis, Günther, Biologia Centrali-Americana, Reptilia, p. 96, May, 1893.

Apatzingán; Coalcomán (6); El Sabino (2).

Most specimens were found beneath rocks in grassy areas near the upper limits of the arid scrub forest, both in the Sierra de Coalcomán and on the southern slopes of the Cordillera Volcánica; all are from elevations of less than 1100 meters. One specimen was found on a road at night near Apatzingán. This species has been found in similar habitats near Huajintlán, Guerrero, and in arid scrub forest at lower elevations in Colima. It is unknown from the coast of Michoacán.

PSEUDOFICIMIA PULCHERRIMA TAYLOR AND SMITH

Pseudoficimia pulcherrima Taylor and Smith, Univ. Kansas Sci. Bull., 28:246, May 15, 1942.—Huajintlán, Guerrero, México.

Apatzingán.

This specimen (CNHM 39208) was reported by Schmidt and Shannon (1947:81); they stated that it was a paratype of *P. pulcherrima*. However, Taylor and Smith (1942a:246) did not mention the specimen; aside from the type (EHT-HMS 5497), the only other specimen they designated as belonging to the type series was UMMZ 85711 from Chilpancingo, Guerrero.

The taxonomic validity of *Pseudoficimia pulcherrima* remains doubtful, for only minor characters distinguish it from *P. frontalis*. Furthermore, all known specimens of *P. pulcherrima* are from within the geographic range of *P. frontalis*.

RHADINAEA HESPERIA HESPERIA BAILEY

Rhadinaea hesperia Bailey, Occ. Pap. Mus. Zool. Univ. Michigan, 412:8, May 6, 1940.—Omilteme and Sierra de Burro, Guerrero. Type locality restricted to Omilteme, Guerrero, México, by Smith and Taylor (1950a:332).

Rhadinaea hesperia hesperia, Smith, Proc. Biol. Soc. Washington, 55:185, December 31, 1942.

Arteaga (3); Coalcomán; El Sabino (2); Uruapan; Volcán Jorullo (2).

One specimen from Volcán Jorullo (UMMZ 104494), three from Arteaga (UMMZ 119281), and one from Uruapan (UMMZ 92342) are typical of the subspecies *R. h. hesperia* in possessing a lateral cream-colored line on the sixth and parts of the fifth and seventh dorsal scale rows and in lacking a dark line on the second scale row. The specimens from El Sabino (EHT-HMS 5441 and UIMNH 18933) and one from Coalcomán (UMMZ 104502) have the cream-colored line on the sixth and adjacent parts of the fifth and seventh scale rows and have a dark line on the second scale row. Another individual from Volcán Jorullo (UMMZ 104682) has cream-colored lines like the others, but it possesses two lateral dark lines, one on the second scale row, and one on the third.

Smith (1942d:186) diagnosed *Rhadinaea hesperia hesperioides* as differing from the nominal subspecies in having the cream-colored line on the fourth and fifth scale rows and in possessing a dark line on the second scale row. The specimens seen all have the lateral cream-colored line centered on the sixth scale row, as is characteristic of *R. h. hesperia*. Although many of the specimens also possess a dark line on the second scale row, these specimens are here assigned to *R. h. hesperia*. Additional specimens are necessary to define accurately the subspecies and their ranges. Peters (1954:29) assigned the specimens from Coalcomán to *R. h. hesperioides*.

In life the specimens from Arteaga had bright cream-colored temporal stripes and dorsolateral stripes on the anterior part of the body. The chin and anterior one-sixth of the belly was white; posteriorly the venter was bright orange-red.

In Michoacán this snake has been found in tropical semi-deciduous forest, arid scrub forest, and pine-oak forest at elevations from 850 to 1500 meters.

RHADINAEA LAUREATA (GÜNTHER)

Dromicus laureatus Günther, Ann. Mag. Nat. Hist., ser. 4, 1:419, 1868.—Mexico City.

Rhadinaea laureata, Boulenger, Catalogue Snakes... British Museum, 2, p. 179, September 23, 1894.

Capácuaro; Carapan (8); Cherán (3); Paracho (2); Pátzcuaro; Tancítaro (10).

This snake is abundant in the Cordillera Volcánica, but it is unknown in the mountains to the northeast of Morelia or in the Sierra de Coalcomán. Most specimens were found beneath volcanic rocks imbedded in the ashy soil in pine forest between 1800 and 2300 meters.

RHADINAEA TAENIATA (PETERS)

Dromicus taeniatus Peters, Monats. Akad. Wiss. Berlin, p. 275, 1863.—México.

Rhadinaea taeniata, Bailey, Occ. Pap. Mus. Zool. Univ. Michigan, 412:14, May 6, 1940.

Tancítaro (2).

This species, which is known only from a small region in the mountains of Jalisco and central Michoacán, is represented by two specimens (CNHM 37130 and 39030) collected at Tancítaro (see Schmidt and Shannon, 1947:80).

SALVADORA BAIRDI JAN

Salvadora Bairdii Jan. Icon. gener. ophid., livr. 2, pl. 3, fig. 2, 1860.—México. Type locality restricted to Acámbaro, Guanajuato, México, by Smith and Taylor (1950a:330).

Barranca Seca; Carapan; Cerro San Andrés; Cojumatlán (2); Jiquilpan; Morelia; Pátzcuaro (4); Quiroga; Sahuayo (2); Tacícuaro (12); Tancítaro (56); Uruapan (2); Zacapu (2); between Zitácuaro and Tuxpan (3).

This species is abundant on the Mexican plateau, where it inhabits the more grassy areas in the mesquite grassland and cutover land in the pine forests from 1550 to 2500 meters. Davis and Dixon (1957:21) described a specimen from Zacapu as having two dark paravertebral stripes diverging on the temporals and extending through the eye onto the loreal, a characteristic of *Salvadora lineata*. On the basis of this specimen, Davis and Dixon suggested that *Salvadora bairdi* and *S. lineata* were subspecifically related. The examination of the large number of specimens from Michoacán has revealed this kind of coloration in only one other specimen, an individual from Tacícuaro, in which the stripes diverge, but do not extend through the eye onto the loreal. Data on scutellation for the large series from Tancítaro were given by Schmidt and Shannon (1947:78), and for the series from Tacícuaro by Smith (1943:466).

SALVADORA MEXICANA (DUMÉRIL, BIBRON, AND DUMÉRIL)

Zamenis mexicanus Duméril, Bibron, and Duméril, Erpétologie genérale, 7 (pt. 1), p. 695, 1854.—Cape Corrientes, Jalisco, México.

Salvadora mexicana Günther, Ann. Mag. Nat. Hist., ser. 3, 12:349, 1863.

Apatzingán (12); Capirio (2); El Sabino (5); Huetamo; La Placita; La Playa (4); Lombardia; Nueva Italia; Ojos de Agua de San Telmo; Oropeo; Río Cancita, 14 km. E of Apatzingán; Santa Ana.

This is one of the most abundant snakes in the arid lowlands of the Tepalcatepec Valley; observations indicate that it probably is equally abundant on the coastal lowlands. Near Apatzingán as many as five of these snakes have been seen in one-half hour. The snakes seem to be equally abundant and active in the dry season and in the rainy season. Most individuals were seen on the ground, but two were found in low trees. On several occasions *Salvadora mexicana* was observed in pursuit of lizards on the ground. Captured individuals regurgitated *Cnemidophorus costatus zweifeli*, *Cnemidophorus deppei infernalis*, *Sceloporus horridus oligoporus*, *Sceloporus pyrocephalus*, and *Urosaurus gadowi*.

Salvadora mexicana inhabits only the arid scrub forest at elevations from sea level to about 1000 meters.

SIBON NEBULATUS (LINNAEUS)

Coluber nebulatus Linnaeus, Systema naturae, ed. 10, 1, p. 222, 1758.—Africa (in error). Type locality restricted to Jicaltepec, Veracruz, México, by Smith and Taylor (1950a:349).

Sibon nebulatus, Taylor, Univ. Kansas Sci. Bull., 26:473, November 27, 1940.

Aquila.

The one specimen from Michoacán was collected by Peters (1954:30) in tropical semi-deciduous forest on the coastal foothills of the Sierra de Coalcomán. As presently known, the range of this species in western México extends from Chiapas to Nayarit. Throughout this region the species avoids scrub forest; this may explain its absence in the Balsas-Tepalcatepec Valley.

SONORA MICHOACANENSIS MICHOACANENSIS (DUGÈS)

Contia michoacanensis Dugès, *in* Cope, Proc. Amer. Philos. Soc., 22:178, 1885.—Michoacán. Type locality restricted to Apatzingán, Michoacán, México, by Smith and Taylor (1950a:335).

Sonora michoacanensis michoacanensis, Stickel, Proc. Biol. Soc. Washington, 56:116, October 19, 1943.

Apatzingán (3); Coalcomán (3); 12 km. S of Tzitzio.

These specimens, together with all known specimens from the Sierra del Sur in Guerrero (KU 23790-1, MVZ 45123) and the upper Balsas Basin in Puebla (UIMNH 41688), are referable to *S. m. michoacanensis*. The dorsal pattern consists of a highly variable number of cross-bands of red, white, and black. In the specimens from Michoacán there are as many as 17 red cross-bands on the body. One specimen from Apatzingán (CNHM 37141) has just behind the head a white band, bordered on either side by a narrow black band; posteriorly the body is uniform red. Two specimens from Coalcomán (UMMZ 109905-6) have respectively 11 and 13 red cross-bands and 20 and 17 white cross-bands, and the posterior part of the body is devoid of red color. Other specimens from these localities have red, black, and white cross-bands throughout the length of the body.

Sonora michoacanensis michoacanensis is distinguished from *S. michoacanensis mutabilis* by the presence of cross-bands on the tail in the latter (Stickel, 1943:116). One specimen from Coalcomán (UMMZ 109904) has one narrow band on the tail; all others from Michoacán have uniformly red tails.

Apparently *Sonora michoacanensis michoacanensis* ranges in semi-arid and arid habitats from the upper Balsas Basin in Puebla westward to the lower slopes of the Sierra de Coalcomán, whereas *S. m. mutabilis* lives in foothills of the Sierra Madre Occidental from southern Jalisco to Nayarit. Zweifel (1959b:6) presented evidence to show that specimens of *S. m. mutabilis* supposedly from "Distrito Federal" probably bear erroneous locality data.

TANTILLA BOCOURTI (GÜNTHER)

Homalocranium bocourti Günther, Biologia Centrali-Americana, Reptilia, p. 149, 1895.—Guanajuato, Guanajuato, México.

Tantilla bocourti, Cope, Amer. Nat., 30:1021, December, 1896.

Carapan; Pátzcuaro (2); between Zitácuaro and Río Tuxpan (11).

This small snake is an inhabitant of the coniferous forests and the pine-oak forests on the Cordillera Volcánica. Data on the series from between Zitácuaro and the Río Tuxpan were given by Taylor (1940c:481).

TANTILLA CALAMARINA COPE

Tantilla calamarina Cope, Proc. Acad. Nat. Sci. Philadelphia, 18:320, February 13, 1867.—Guadalajara, Jalisco, México.

Apatzingán; La Placita.

Although this snake has been collected at high elevations along the rim of the Mexican Plateau in Nayarit, Jalisco, México, and Puebla, the specimens from Michoacán are from arid scrub forest at elevations of less than 400 meters. The species has been found in similar habitats in Colima (Oliver, 1937:24) and in Sinaloa and the Tres Marías Islands (Zweifel, 1960:110).

TOLUCA LINEATA LINEATA KENNICOTT

Toluca lineata Kennicott, *in* Baird, Report on the United States and Mexican boundary survey, 2, Reptiles, p. 23, 1859.—Valley of México.

Toluca lineata lineata, Taylor and Smith, Univ. Kansas Sci. Bull., 28:343, May 15, 1942.

Capácuaro; Carapan (12); Cherán (23); Cojumatlán; Los Reyes; Morelia (2); Nahuatzen; Paracho (10); Pátzcuaro (17); Uruapan (2).

This small snake is an inhabitant of the coniferous forests between elevations of about 1550 and 2800 meters. Not infrequently, individuals have been found in pine-oak forest within these elevations.

The generic status of *Toluca* is unsettled. Taylor and Smith (1942b) separated *Toluca* from *Conopsis* by the presence of enlarged and grooved posterior maxillary teeth in *Toluca* and their absence in *Conopsis*. Bogert and Oliver (1945:378) suggested synonymizing *Toluca* with *Conopsis*. Smith and Laufe (1945:12) defined the generic position of *Toluca*. Actually, in deciding the generic position of these snakes, five genera (*Ficimia*, *Gyalopion*, *Pseudoficimia*, *Conopsis*, and *Toluca*) must be considered. Of these *Ficimia* and *Gyalopion* are closely related; they have been placed in one genus by some workers. *Pseudoficimia* is intermediate between *Ficimia-Gyalopion* and *Toluca-Conopsis*. A workable definition of the supraspecific classification of these snakes must await a thorough review of the species.

TRIMORPHODON BISCUTATUS BISCUTATUS (DUMÉRIL, BIBRON, AND DUMÉRIL)

Dipsas biscutata Duméril, Bibron, and Duméril, Erpétologie genérale, 7 (pt. 2):1153, 1854.—México. Type locality restricted to Tehuantepec, Oaxaca, México, by Smith and Taylor (1950a:340).

Trimorphodon biscutatus biscutatus, Smith, Proc. U. S. Natl. Mus., 91:159, November 10, 1941.

Apatzingán (11); Cofradía; Cuatro Caminos; El Sabino (2); La Placita; La Playa (2); Lombardia (2); Nueva Italia (2); Río Tepalcatepec, 27 km. S of Apatzingán; Tafetán.

In the arid lowlands of the Tepalcatepec Valley and presumably also in the scrub forest of the coastal lowlands, this is an abundant snake, which is active only at night. Usually snakes of this species are found on the ground, but one large individual was observed at night in a low tree. That individual defied capture by widely opening its mouth and striking repeatedly at the collector. The excreta of one specimen contained feathers of an unidentified species of bird.

TRIMORPHODON LATIFASCIA PETERS

Trimorphodon biscutata latifascia Peters, Monats. Akad. Wiss. Berlin, p. 877, 1869.—Puebla, México. Type locality restricted to Izúcar de Matamoros, Puebla, México, by Smith and Taylor (1950a:341).

Trimorphodon latifascia. Taylor, Univ. Kansas Sci. Bull., 25:364, July 10, 1939.

Apatzingán (5); Casada Tzararacua; Coalcomán (2); Lombardia; 14 km. S of Lombardia; Nueva Italia; San Salvador.

In Michoacán this species has been collected in semi-arid habitats at elevations from 300 to 1430 meters in the Tepalcatepec Valley and lower slopes of the Cordillera Volcánica. In this area it occurs sympatrically with *Trimorphodon biscutatus biscutatus*.

In life, adults have a pale tan dorsal ground color and rich chocolate brown cross-bands; the eye is pale grayish tan. A juvenile from Coalcomán has black cross-bands on a pale grayish tan ground color. As stated by Schmidt and Shannon (1947:83) and Peters (1954:32), the type specimen of *Trimorphodon fasciolata* Smith from Cascada Tzararacua is indistinguishable from specimens of *Trimorphodon latifascia*.

Seven males have 209 to 223 (216.5) ventrals; one female has 227 ventrals. The number of dark cross-bands on the body varies from 12 to 16 (13.5). The relationships of this species are with *Trimorphodon tau* on the Mexican Plateau. In fact, additional specimens from the headwaters of the Tepalcatepec Valley and the lower slopes of the Mexican Plateau in eastern Michoacán and adjacent Jalisco may show that the two are conspecific. *Trimorphodon latifascia* differs from *tau* in having fewer dark cross-bands on the body and in lacking an interocular bar.

TRIMORPHODON TAU COPE

Trimorphodon tau Cope, Proc. Amer. Philos. Soc., 11:151, 1869.—Quiotepec, Oaxaca, México.

Emiliano Zapata (2); between Morelia and Ciudad Hidalgo; Tacícuaro; Tangamandapio.

Two of the specimens from Michoacán (UMMZ 118948 from Tangamandapio and UIMNH 19138 from Tacícuaro) have cream-colored, Y-shaped marks on the head. These markings supposedly are characteristic of *Trimorphodon upsilon*. One specimen from Emiliano Zapata (UMMZ 118950) and one from between Morelia and Ciudad Hidalgo (EHT-HMS 21402) have a cream-colored line on the parietal suture; in another specimen from Emiliano Zapata (UMMZ 118949) the anterior end of this line is expanded, giving the appearance of an incipient "Y". Thus, the nature of the markings on the head in specimens from Michoacán is intermediate between the typical condition in *Trimorphodon tau* and the usual condition in *T. upsilon*. Smith and Taylor (1945:148) gave the range of *Trimorphodon tau* as: "Central Guerrero, in the Sierra Madre del Sur; central Oaxaca; and the edge of the plateau in central Michoacán." They gave the range of *Trimorphodon upsilon* as: "Southern Chihuahua south to central Michoacán, east to central Hidalgo." Specimens referable to *T. tau* have been found at La Joya de Salas, near Ciudad Victoria, and near Llera, Tamaulipas (see Smith and Darling, 1952:85, and Martin, 1958:74). Some of these specimens also show combinations of characteristics of *T. tau* and *T. upsilon*. Smith and Darling (*loc. cit.*) suggested that *T. tau* and *T. upsilon* be considered as subspecies. However, if *T. tau* and *T. upsilon* are subspecies, intergrades would be expected between the ranges of the two populations and not on the northeastern and southwestern periphery of their combined ranges. Instead, the limited evidence now available suggests that *T. tau* and *T. upsilon* are names based on a highly variable character of color pattern of the head, and that only one species is involved.

In Michoacán this species inhabits the mesquite grassland on the Mexican Plateau.

TROPIDODIPSAS OCCIDENTALA OLIVER

Tropidodipsas occidentala Oliver, Occ. Pap. Mus. Zool. Univ. Michigan, 360:20, November 20, 1937.—Comala, Colima, México.

Coalcomán.

This specimen was reported by Peters (1954:34), who found it beneath a rock at the mouth of a heavily wooded ravine near Coalcomán at an elevation of 950 meters. The only other known specimen is from Comala, Colima, a village, like Coalcomán, that is located near the upper limits of the arid scrub forest.

NATRIX VALIDA ISABELLEAE CONANT

Natrix valida isabelleae Conant, Nat. Hist. Misc., 126:7, September 15, 1953.—Pie de la Cuesta, Laguna Coyuca, Guerrero, México.

Coahuayana; Playa Azul (2); Punto San Juan de Lima.

Three females and one male have, respectively, 133, 135, 135, and 131 ventrals, and 68, 68, 73, and 75 caudals. The grayish stippling on the posterior ventral surfaces mentioned by Conant (1953:9) is not visible on these specimens. In the small individuals from Punto San Juan de Lima and from Coahuayana there are four longitudinal rows of dark spots on the dorsum; in two large females from Playa Azul the spots are barely discernible.

All of the specimens from Michoacán were found in the coastal lowlands; those from Playa Azul were collected from a small brackish, mangrove-lined lagoon.

STORERIA STORERIOIDES (COPE)

Tropidoclonium storerioides Cope, Proc. Acad. Nat. Sci. Philadelphia, 17:190, December 26, 1865.—Mexican Plateau. Type locality restricted to Tres Cumbres, Morelos, México, by Smith and Taylor (1950a:336).

Storeria storerioides, Garman, Mem. Mus. Comp. Zool., 8(3):29, June, 1883.

Dos Aguas (11); Puerto de Garnica; Tancítaro (11); Tzitzio; Uruapan; 16 km. NW of Zacapu.

Three males and six females from the Sierra de Coalcomán have, respectively, 122-128 (125.3) and 126-136 (130.0) ventrals, and 46-47 (46.7) and 38-42 (39.1) caudals. Four males and eleven females from the Cordillera Volcánica have, respectively, 124-132 (128.5) and 127-139 (136.4) ventrals, and 43-48 (44.7) and 38-44 (40.2) caudals. These data show that, although there is little difference in the number of caudals, specimens from the Sierra de Coalcomán have fewer ventrals than do specimens from the Cordillera Volcánica. Of eleven specimens from the Sierra de Coalcomán, two have black bellies. Five others from the Sierra de Coalcomán and one from Puerto de Garnica in the Cordillera Volcánica have the bellies heavily stippled with black, giving a gray appearance. Melanistic tendencies in this species have been discussed by Anderson (1960:64), who examined the specimen from Tzitzio. In life, one specimen from Dos Aguas (UMMZ 119451) had a cream-colored belly; the edges of the ventrals were dark brick-red.

In Michoacán this snake inhabits pine-oak, pine, and fir forests at elevations between 1550 and 2800 meters in the Cordillera Volcánica and the Sierra de Coalcomán. Most specimens were found beneath rocks; the one from Tzitzio was removed from the stomach of a Mexican Motmot (Anderson, 1960:66).

THAMNOPHIS DORSALIS CYCLIDES COPE

Thamnophis cyrtopsis cyclides Cope, Proc. Acad. Nat. Sci. Philadelphia, 13:299, December 28, 1861.—Cape San Lucas, Baja California (in error). Type locality restricted to Guanajuato, Guanajuato, México, by Smith and Taylor (1950a:330). Smith, Copeia, no. 2:140, June 8, 1951. Milstead, Texas Jour. Sci., 5:368, September, 1953.

Thamnophis eques eques (*nec.* Reuss), Smith, Zoologica, 27:106, October 23, 1942. Bogert and Oliver, Bull. Amer. Mus. Nat. Hist., 83:356, March 30, 1945.

Thamnophis vicinus Smith, Zoologica, 27:104, October 23, 1942.—Temazcal, Michoacán, México.

Thamnophis dorsalis cyclides, Fitch and Milstead, Copeia, no. 1:112, March 17, 1961.

Barolosa; Coalcomán; Dos Aguas (3); Los Reyes; Morelia (16); Opopeo; Pino Gordo; Tacícuaro (16); Tancítaro (14); Tangamandapio (2); Temazcal (2); Tzintzuntzan; Uruapan.

The snakes comprising the former *Thamnophis eques*-group have undergone extensive taxonomic and nomenclatural shuffling by Smith (1942 and 1951), Bogert and Oliver (1945), Milstead (1953), and Fitch and Milstead (1961). Smith recognized in Michoacán three members of the *T. eques* (= *dorsalis*) complex: *eques eques*, *eques postremus*, and *vicinus*. Later, Smith (1951) showed that the specific name *eques* had been misapplied, so that *T. eques eques* became *T. cyrtopsis cyclides*, and *T. eques postremus* became *T. cyrtopsis postremus*; under this arrangement *T. vicinus* stood unchanged. In the meantime, Bogert and Oliver (1945:359) presented a reinterpretation of Smith's data and suggested that *T. vicinus*, which differs from *T. dorsalis cyclides* only in lacking a middorsal stripe, "... is not a species, but only a pattern phase, possibly a simple mutant of *T. e. eques*" (= *T. dorsalis cyclides*, by present arrangement). Milstead (1953) agreed with Bogert and Oliver on the status of *T. vicinus*; furthermore, on the basis of only a few specimens, Milstead concluded that *T. cyrtopsis postremus* was not subspecifically distinct from *T. cyrtopsis cyclides*. Recently, Fitch and Milstead (1961) showed that *Thamnophis dorsalis* Baird and Girard (1853) was the correct name for the snakes that had been recognized as *Thamnophis cyrtopsis* Kennicott (1860). Consequently, the snakes referred to *T. eques eques* by Smith (1942) and to *T. cyrtopsis cyclides* by Smith (1951) and Milstead (1953) are now *T. dorsalis cyclides*.

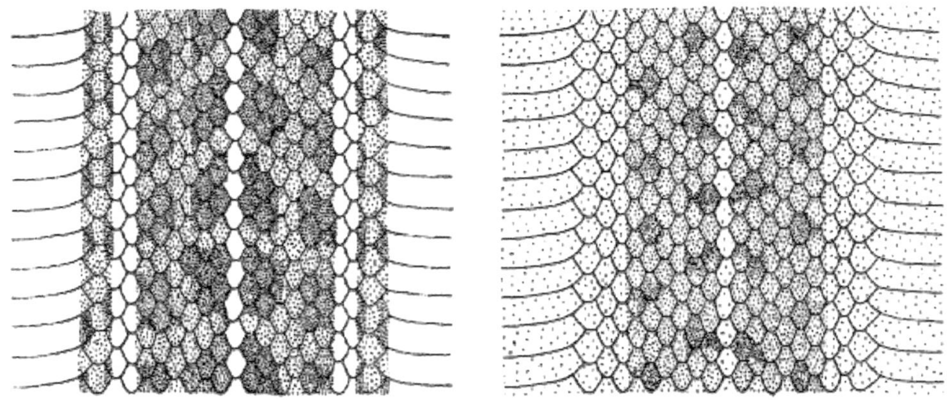

Fig. 10. Dorsal color pattern of Thamnophis dorsalis cyclides (A) and Thamnophis dorsalis postremus (B).

Aside from one specimen from Temazcal and nine from Morelia (paratypes of *T. vicinus*), only two other specimens completely lacking the middorsal stripe have been seen; one is a male (UMMZ 102510) having 161 ventrals and an incomplete tail from Pino Gordo, and the other is a male (CNHM 39060) from Tancítaro having 158 ventrals and an incomplete tail. A female from Tancítaro (CNHM 39061) having 153 ventrals and 77 caudals has no lateral stripes and only a narrow middorsal stripe on the anterior part of the body. Throughout the region where *T. vicinus*-like snakes have been found, typical *T. dorsalis cyclides* occurs in much greater numbers. I concur with Bogert and Oliver in placing *T. vicinus* as a synonym of *T. dorsalis cyclides*.

Milstead (1953) had available few specimens of *Thamnophis dorsalis* from the Tepalcatepec Valley. The large series now in existence shows that the population in the Tepalcatepec Valley differs distinctly from that inhabiting the Mexican Plateau, Cordillera Volcánica, and Sierra de Coalcomán. Therefore the name *T. dorsalis postremus* Smith (1942) is resurrected for the population in the Tepalcatepec Valley. *T. dorsalis cyclides* and *T. dorsalis postremus* differ in color pattern (Fig. 10) and in scutellation (Table 6). Specimens from the Mexican Plateau and mountain ranges have a distinct light stripe on the second and third scale rows, a dark brown dorsum having squarish black spots, and a row of dark spots on the first row of dorsal scales. Specimens from the Tepalcatepec Valley have a grayish brown dorsum having smaller and less distinct dark spots and no light stripe on the second and third scale rows; the first, second, and third rows of scales are colored like the venter. In some specimens there are small dark flecks on the first row of dorsal scales.

TABLE 6.— VARIATION IN SCUTELLATION IN THAMNOPHIS DORSALIS.

Character		Mexican Plateau	Sierra de Coalcomán	Tepalcatepec Valley
Ventrals	♀ N	31	2	32
	Mean	164.0	156.5	144.6
	Range	153-171	154-159	138-151
	♂ N	19	2	32
	Mean	153.5	154.7	138.3
	Range	149-159	149-159	131-141
Caudals	♀ N	28	2	29
	Mean	83.8	81.0	73.4
	Range	80-100	79-83	70-79
	♂ N	14	2	28
	Mean	78.0	72.0	68.5
	Range	71-87	72	63-73

One specimen from Uruapan (1550 meters) and one from Coalcomán (950 meters) are intermediate in color pattern between *T. dorsalis cyclides* and *T. dorsalis postremus*. Both have indistinct lateral stripes and only small dark spots below the stripes. In scutellation these specimens are like *T. dorsalis cyclides*.

In Michoacán *Thamnophis dorsalis cyclides* has been collected in a variety of habitats on the Mexican Plateau: pine-oak forest, fir forest, marshes, and cleared land from 1550 to 2800 meters. In the Sierra de Coalcomán one was taken in broadleafed forest at 950 meters, three in pine-oak forest at 2100 meters, and one in pine forest at 2300 meters.

THAMNOPHIS DORSALIS POSTREMUS SMITH

Thamnophis eques postremus Smith, Zoologica, 27:109, October 23, 1942.—El Sabino, Michoacán, México.

Thamnophis cyrtopsis postremus Smith, Copeia, no. 2:140, June 8, 1951.

Thamnophis cyrtopsis cyclides (part), Milstead, Texas Jour. Sci., 5:368, September, 1953.

Thamnophis dorsalis postremus, Fitch and Milstead, Copeia, no. 1:112, March 17, 1961.

Apatzingán (31); Capirio (2); Charapendo; Cuatro Caminos (22); El Sabino; Lombardia (9); Nueva Italia (8); Uruapan (3).

The reasons for recognizing the population of *Thamnophis dorsalis* in the Tepalcatepec Valley as distinct from that on the surrounding highlands are presented in the discussion of *Thamnophis dorsalis cyclides*. In certain features of coloration and in the low numbers of ventrals and caudals, *T. dorsalis postremus* shows more resemblance to *T. dorsalis sumichrasti* than to *T. dorsalis cyclides*. According to Milstead (1953:367), *T. dorsalis cyclides* ranges southward from the Río Balsas in southwestern México. If specimens could be obtained from the upper Balsas Basin they might show that *T. dorsalis postremus* inhabits that extensive basin.

In the Tepalcatepec Valley *T. dorsalis postremus* is most frequently found at night in the rainy season, at which time the snakes are abundant near temporary pools where frogs are breeding. The absence of specimens from the coastal lowlands of Guerrero, Michoacán, and Colima indicate that, although the species inhabits the lowlands of the Tepalcatepec Valley, its range does not include the coastal lowlands.

A female (UMMZ 119402 from Cuatro Caminos) having 139 ventrals and a body length of 576 mm., on June 20, 1958, gave birth to 25 young, of which 18 (9 males and 9 females) were preserved. In body length the males varied from 132 to 141 (137.3) mm.; the females, 125 to 137 (133.1) mm. In tail length the males varied from 38 to 44 (42.4) mm.; females, 35 to 42 (39.7) mm. The males have 138 to 147 (142.2) ventrals and 70 to 75 (72.9) caudals; females have 131 to 140 (135.8) ventrals and 63 to 71 (67.0) caudals.

THAMNOPHIS EQUES EQUES (REUSS)

Coluber eques Reuss, Zool. Misc., p. 152, 1834.—México. Type locality restricted to Guadalajara, Jalisco, México, by Smith and Taylor (1950a:334).

Thamnophis macrostemma macrostemma, Smith and Taylor, Bull. U. S. Natl. Mus., 187:163, October 5, 1945.

Thamnophis subcarinata subcarinata, Smith, Herpetologica, 5:63, May 31, 1949.

Thamnophis eques eques, Smith, Copeia, no. 2:139, June 8, 1951.

Jiquilpan; Lago de Cuitzeo; Lago de Pátzcuaro (17); Pátzcuaro (5); Tangancícuaro; Tupátaro (2); Undameo; Zacapu.

Although this snake has been collected in open pine-oak forest and in oak-bunch grass associations, it seems to reach its greatest abundance in marshes on the Mexican Plateau at elevations of 1550 to 2300 meters.

THAMNOPHIS MELANOGASTER CANESCENS SMITH

Thamnophis melanogaster canescens Smith, Zoologica, 27:117, October 23, 1942.—Chapala, Jalisco, México.

Lago de Cuitzeo (5); Lago de Pátzcuaro; Pátzcuaro; Tacícuaro; Tangamandapio (2).

This species of garter snake seems to be most abundant in the marshes adjacent to the lakes on the Mexican Plateau in Michoacán and Jalisco. At these elevations (1550 to 2200 meters) it often is found in association with *Thamnophis eques eques* and sometimes with *Thamnophis dorsalis cyclides*. On June 11, 1958, individuals of this species were found in a hyacinth-choked marsh at Tangamandapio at night.

One specimen from Tangamandapio (UMMZ 119414) had, in life, a dark chocolate brown dorsum, reddish brown sides, and cream-colored belly, chin, and labials. There were no longitudinal dorsal stripes.

THAMNOPHIS SCALARIS SCALIGER (JAN)

Tropidonotus scaliger Jan, Elenco sistematico degli Ofidi, p. 70, 1863.—No type locality designated. Type locality restricted to Mexico City, Distrito Federal, by Smith and Taylor (1950a:329).

Thamnophis scalaris scaliger, Smith, Zoologica, 27:103, October 23, 1942.

Cerro Tancítaro (2); Nahuatzen; Opopeo; 26 km. S of Pátzcuaro.

The few specimens of this species from Michoacán have been collected at elevations from 1800 to 3400 meters in pine or fir forest in the Cordillera Volcánica.

MICRURUS DISTANS MICHOACANENSIS (DUGÈS)

Elaps diastema michoacanensis Dugès, La Naturaleza, ser. 2, 1:487, 1891.—Michoacán. Type locality restricted to Apatzingán, Michoacán, México, by Smith and Taylor (1950a:335).

Micrurus distans michoacanensis, Zweifel, Amer. Mus. Novitates, 1953:11, June 26, 1959.

Apatzingán (6).

All specimens were collected in the arid scrub forest of the Tepalcatepec Valley. The number of black rings on the body varies from six to eleven. In this respect they agree with the diagnosis of this subspecies presented by Zweifel (1959b:9).

MICRURUS LATICOLLARIS (PETERS)

Elaps marcgravii laticollaris Peters, Monats. Akad. Wiss. Berlin, p. 877, 1869.—Izúcar de Matamoros, Puebla, México.

Micrurus laticollaris, Schmidt, Publ. Field Mus. Nat. Hist., zool. ser., 20:39, December 11, 1933.

El Sabino (2); Lombardia.

This species ranges throughout the Balsas-Tepalcatepec Basin westward into Colima; specimens from Michoacán were collected in arid scrub forest at elevations from 500 to 1050 meters. The limited observations on *Micrurus distans michoacanensis* and *M. laticollaris* indicate that, at least in the Tepalcatepec Valley, *M. laticollaris* seems to inhabit slightly more mesic areas than does *M. distans michoacanensis*.

PELAMIS PLATURUS (LINNAEUS)

Anguis platura Linnaeus, Systema naturae, ed. 12, 1:391, 1766.—Pine Island, Pacific Ocean.

Pelamis platurus, Gray, Ann. Philos., p. 15, 1825.

Boca de Apiza.

In November, 1955, Alfonzo Gonzales, a geographer from the University of Texas, observed sea snakes on the beaches of Michoacán. In May, 1956, Donald D. Brand of the University of Texas gave me one specimen of *Pelamis platurus* that he obtained on March 2, 1956, at Boca de Apiza. Furthermore, he supplied me with the following observations based on his field work along the coast of Michoacán from the Río Coahuayana to Maruata from March 1, to April 15, 1956. At that time many sea snakes were observed; in some places living and dead individuals were seen on the beaches; innumerable snakes were seen in the surf. When live individuals were taken from the beach and thrown into the ocean, they usually swam to shore. Many partially eaten individuals were seen protruding from crab holes. Inquiries among the natives resulted in the following information: Sea snakes are frequently seen between November and April, but most commonly in March and April, at which time the water is cold. The natives referred to the sea snakes as "culebra del mar." Most natives said that the snakes were not poisonous; others did not know of any venomous properties. In May, 1956, I worked the coastal region from the Río Coahuayana to La Placita and saw no sea snakes. In the summer of 1950 James A. Peters, and in the summer of 1951 I worked nearly the entire coastal region of Michoacán; during that time no *Pelamis* were seen. Insofar as I know, this is the first report of such seasonal activity in *Pelamis platurus* in the Americas.

AGKISTRODON BILINEATUS BILINEATUS GÜNTHER

Ancisdrodon bilineatus Günther, Ann. Mag. Nat. Hist., ser. 3, 12:364, 1863.—Pacific coast of Guatemala.

Agkistrodon bilineatus bilineatus, Burger and Robertson, Univ. Kansas Sci. Bull., 34 (1):213, October 1, 1951.

Apatzingán; El Sabino; La Playa; Los Reyes.

All specimens from Michoacán are from inland localities between 300 and 1500 meters. The one from Los Reyes (USNM 46416) was collected by Nelson and Goldman on February 13, 1903. The elevation of Los Reyes (1500 meters) seems unusually high for this species, but otherwise there is no reason to doubt the authenticity of the record. Goldman (1951:192) in his description of Los Reyes stated: "Los Reyes is near the boundary between the Lower Austral and Arid Upper Tropical Zones but is preponderantly tropical in zonal character. The regular crops are mainly sugar cane, rice, and corn." Thus the biotic features of the area are not noticeably different from those at El Sabino and La Playa at lower elevations. The development of extensive agriculture through irrigation in the Tepalcatepec Valley and planting of rice and sugar-cane in that area may produce a more widespread habitat for this snake.

The absence of specimens from the coastal lowlands is due solely to inadequate collecting; the natives there know the snake and report that it is not uncommon in certain areas.

CROTALUS BASILISCUS BASILISCUS (COPE)

Caudisonia basilisca Cope, Proc. Acad. Nat. Sci. Philadelphia, September 30, 1864.—Colima. Type locality restricted to Colima, Colima, México, by Smith and Taylor (1950a:328).

Crotalus basiliscus basiliscus, Gloyd, Nat. Hist. Misc., 17:1, April 23, 1948.

Apatzingán (4); Camachines; Coalcomán; El Ticuiz.

Specimens from southern Michoacán have fewer ventrals and caudals than do those from the northern part of the range; three males and three females have, respectively, 178, 182, 182, 185, 186, and 188 ventrals, and 27, 28, 29, 22, 29, and 29 caudals. Klauber (1952:81) gave the following data for *Crotalus basiliscus* (based on specimens from the entire range, except Oaxaca): ventrals in males, 179-201 (191.4), in females, 185-206 (197.6); caudals in males, 26-36 (30.7), in females, 21-29 (24.4). Klauber (1952:84) remarked that the one specimen that he had seen from Apatzingán had fewer ventrals and caudals than most other specimens. The low numbers of ventrals and caudals in specimens from Michoacán, as compared with more northern populations, may be indicative of a trend in the reduction of the numbers of these scutes from north to south. The southernmost examples of *Crotalus basiliscus* (*Crotalus basiliscus oaxacus* from Oaxaca) have 172-175 ventrals and 21 caudals (Gloyd, 1948).

In Michoacán *Crotalus basiliscus basiliscus* has been found in arid habitats on the coast, in the Tepalcatepec Valley, and in the lower parts of the Sierra de Coalcomán. All specimens are from localities below 1070 meters in elevation.

CROTALUS DURISSUS CULMINATUS KLAUBER

Crotalus durissus culminatus Klauber, Bull. Zool. Soc. San Diego, 26:65, August 8, 1952.—El Sabino, Michoacán, México.

El Sabino (18).

These specimens are part of the type series and were collected by Hobart M. Smith near the upper limits of the arid scrub forest at an elevation of about 1050 meters on the lower slopes of the Cordillera Volcánica at the northern edge of the Tepalcatepec Valley. They were discussed in detail by Klauber (1952:66-70).

CROTALUS INTERMEDIUS INTERMEDIUS TROSCHEL

Crotalus intermedius Troschel, *in* von Müller, Reisen in Vereiningten Staaten, Canada und Mexico, vol. 3, p. 613, 1865.—Type locality unknown.

Crotalus intermedius intermedius, Klauber, Bull. Zool. Soc. San Diego, 26:9, August 8, 1952.

Cerro Tancítaro.

The one specimen is from the pine forests on the Cordillera Volcánica. At the present time this species is known from scattered localities in west-central Veracruz, Oaxaca, Michoacán, and as *Crotalus intermedius omiltemanus* in Central Guerrero. Apparently it is restricted to montane environments.

CROTALUS MOLOSSUS NIGRESCENS GLOYD

Crotalus molossus nigrescens Gloyd, Occ. Pap. Mus. Zool. Univ. Michigan, 325:2, January 28, 1936.—Four miles west of La Colorada, Zacatecas, México.

Carapan; Los Conejos; Pátzcuaro; Tacícuaro (5).

In Michoacán this species has been found in pine forests between 1550 and 2300 meters in the Cordillera Volcánica. I expected to find it in the Sierra de Coalcomán, but inquiries among the natives living in the pine forests of that mountain range revealed that the people there have no knowledge of a large species of rattlesnake.

CROTALUS POLYSTICTUS (COPE)

Caudisonia polysticta Cope, Proc. Acad. Nat. Sci. Philadelphia, 17:191, December 26, 1865.—Tableland of México. Type locality restricted to Tupátaro, Guanajuata, México, by Smith and Taylor (1950a:330).

Crotalus polystictus Cope, *in* Yarrow, Wheeler's Rept. Geog. Geol. Expl. Surv. W. 100th. Mer., vol. 5, p. 533, 1875.

Tacícuaro (4); Tupátaro (2).

Formerly this species was abundant in the marshes around Lago de Chapala. The draining of these marshes probably resulted in reducing the numbers of these rattlesnakes. The species is known only from the Mexican Plateau at elevations of 1450 to 2400 meters.

CROTALUS PUSILLUS KLAUBER

Crotalus pusillus Klauber, Bull. Zool. Soc. San Diego, 26:34, August 8, 1952.—Tancítaro, Michoacán, México.

Acuaro de las Lleguas (2); Carapan; Cerro Tancítaro (16); Dos Aguas (12).

Aside from the type series of *Crotalus pusillus* from Cerro Tancítaro and one specimen from Carapan referred to the species by Klauber (1952:38), there are fourteen specimens from the Sierra de Coalcomán. These specimens (UMMZ 112566-7, 118591-9, 118601, 121512-3) are like *Crotalus pusillus* from Cerro Tancítaro in having the prefrontals paired, a black proximal rattle, and the underside of the tail black. The prefrontals are bordered posteriorly by one scale in two specimens, by two scales in three specimens, and by three scales in the other nine. The snakes from the Sierra de Coalcomán have 40 to 46 (42) dorsal body blotches. Ten males have 150-158 (154.4) ventrals and 29-33 (31.0) caudals; two females have 157 and 160 ventrals, and 25 and 27 caudals. The largest specimen is a male having a body length of 545 mm. and a tail length of 63 mm. The only noticeable difference between the specimens from the Sierra de Coalcomán and the topotypic series is that the latter have fewer dorsal blotches; the range of variation is 33 to 46 (39.8).

Most specimens of this species have a grayish brown dorsum and dark brown dorsal blotches. Two specimens from Dos Aguas (UMMZ 118596 and 118599) are pale brown above and have indistinct blotches.

One specimen from Dos Aguas regurgitated a large *Gerrhonotus imbricatus imbricatus*; of two others from the same locality, one regurgitated a *Sceloporus bulleri* and an *Eptesicus fuscus*. The latter specimen was collected at the entrance of a small cave, where it probably had captured the bat.

In the Cordillera Volcánica *Crotalus pusillus* has been obtained in pine-oak forest at elevations between 1550 and 1800 meters. In the Sierra de Coalcomán two specimens were taken in pine forest at an elevation of 2300 meters; ten other were found beneath rocks and logs in pine-oak forest at an elevation of 2100 meters.

CROTALUS TRISERIATUS AQUILUS KLAUBER

Crotalus triseriatus aquilus Klauber, Bull. Zool. Soc. San Diego, 26:24, August 8, 1952.—Alvarez, San Luis Potosí, México.

Morelia (10); Tacícuaro (2).

I am following Klauber (1952) in assigning some of the specimens of this species from Michoacán to the subspecies *aquilus* and others to *C. t. triseriatus*. The distinguishing characters of these subspecies are given by Klauber (1952:28). On the basis of the few localities from which the species is known in Michoacán it seems as though *C. t. aquilus* inhabits the open grassy areas on the Mexican Plateau and the associated open pine-oak or oak-bunch grass habitats to the north and east of the Cordillera Volcánica. *Crotalus triseriatus aquilus* has been collected at elevations from 1600 to 2000 meters in Michoacán.

CROTALUS TRISERIATUS TRISERIATUS (WAGLER)

Uropsophus triseriatus Wagler, Natürliches System der Amphibien, p. 176, 1830.—México. (Probably Mexico City.)

Crotalus triseriatus triseriatus, Klauber, Bull. Zool. Soc. San Diego, 26:19, August 8, 1952.

Cerro Tancítaro (36); Opopeo; Pátzcuaro.

This small rattlesnake inhabits rocky areas in pine and pine-oak forests above 1600 meters in the Cordillera Volcánica; it has been collected at 3270 meters on Cerro Tancítaro. The series reported by Schmidt and Shannon (1947:84) is a mixture of specimens of *Crotalus triseriatus* and *Crotalus pusillus.* The two species are found together on Cerro Tancítaro, but only *Crotalus pusillus* inhabits the coniferous forests of the Sierra de Coalcomán. Klauber (1952:30) stated that despite the proximity of *Crotalus triseriatus triseriatus* and *Crotalus triseriatus aquilus* in Michoacán, there is no evidence of intergradation. He went on to suggest that additional material might show that the two named populations actually are distinct species. The specimens that have been studied since Klauber's investigations also show no evidence of intergradation, but there still is no known sympatry of the populations.

The small montane rattlesnakes belonging to the species *C. pricei, C. pusillus,* and *C. triseriatus* present a problem in systematics and distribution worthy of intensive investigation. A knowledge of the distribution and relationships of the various populations of these snakes, together with other species also living in isolated populations on the higher mountains in México, probably will be of great significance in understanding dispersal and differentiation of animals during the Pleistocene.

SPECIES OF QUESTIONABLE OCCURRENCE

Some species for which there are no authentic records from Michoacán can be expected there on zoogeographic probability. Other species have been recorded from Michoacán, but these records are doubtful for any one of several reasons. Fifteen species of such questionable occurrence are discussed below:

SYRRHOPHUS MODESTUS MODESTUS TAYLOR

Syrrhophus modestus Taylor, Univ. Kansas Sci. Bull., 28:304, May 15, 1942.—Hacienda Paso del Río, Colima, México.

Syrrhophus modestus modestus, Duellman, Occ. Pap. Mus. Zool. Univ. Michigan, 594:5, June 6, 1958.

This small terrestrial frog is not uncommon on the coastal lowlands and foothills in Nayarit and in Colima, where it has been collected within a few kilometers of the Michoacán border. At Tecolapa, Colima, on August 9, 1956, *Syrrhophus modestus modestus* was found with *Tomodactylus nitidus orarius, Bufo marinus, Bufo marmoreus, Hyla baudini, Hyla smithi*, and *Phyllomedusa dacnicolor*, all of which occur on the coastal lowlands of Michoacán. Because of its solitary and secretive habits, *Syrrhophus modestus modestus* is not common in collections. Additional field work on the coast of Michoacán should reveal the presence of the species there.

HYLA MICROCEPHALA SARTORI SMITH

Hyla microcephala sartori Smith, Herpetologica, 7:186, December 31, 1951.—1 mi. N of Organos, S of El Triente, Guerrero, México.

On August 28, 1960, J. R. Dixon obtained a series of this species from a temporary pond 6 kilometers northeast of La Resolana, Jalisco. Previously, *Hyla microcephala sartori* had been known only from the lowlands of Guerrero and Oaxaca. The existence of the species in Jalisco provides evidence that this frog also occurs in Michoacán and Colima.

GASTROPHRYNE USTA USTA (COPE)

Engystoma ustum Cope, Proc. Acad. Nat. Sci. Philadelphia, 18:131, 1866.—Guadalajara, Jalisco, México.

Gastrophryne usta usta, Carvalho, Occ. Pap. Mus. Zool. Univ. Michigan, 555:13, July 16, 1954.

Smith and Taylor (1948:93-4) listed specimens of this species from Organos and El Treinta, Guerrero, and from Paso del Río, Quesería, Santiago, and Tecomán, Colima. The species occurs from Sinaloa and central Veracruz southward at low elevations to the Isthmus of Tehuantepec and thence along the Pacific lowlands into Central America. Almost certainly it occurs on the coastal lowlands in Michoacán. Since the amphibian fauna of the Tepalcatepec Valley has been better sampled than that of the coast, I suspect that if *Gastrophryne* occurred in the Tepalcatepec Valley, I would have found it there.

LEPIDOCHELYS OLIVACEA (ESCHSCHOLTZ)

Chelonia olivacea Eschscholtz, Zool. Atlas, pt. 1, p. 2, 1829.—Manila Bay, Philippine Islands.

Lepidochelys olivacea, Girard, United States Exploring Expedition..., vol. 20, Herpetology, p. 435, 1858.

According to Smith and Taylor (1950b: 15), this sea turtle is known from the entire Pacific coast of México; these authors reported the species from Chiapas, Oaxaca, Guerrero, Colima, and Sonora. Although the only sea turtle that I observed in Michoacán is *Chelonia mydas*, others probably do use the sheltered beaches for nesting. The scanty records of sea turtles along the Pacific coast of México indicate that *Chelonia mydas* and *Lepidochelys olivacea* are the most abundant species in that region. There are scattered records of *Dermochelys coriacea*, *Caretta caretta*, and *Eretmochelys imbricata* along the Pacific coast. The occurrence of any of these along the coast of Michoacán is probable.

GEOEMYDA PULCHERRIMA PULCHERRIMA (GRAY)

Emys pulcherrima Gray, Catalogue of the Shield Reptiles in British Museum, vol. 1, p. 25.—México. Type locality restricted to Presidio de Mazatlán, Sinaloa, México, by Smith and Taylor (1950b:30).

Geoemyda pulcherrima pulcherrima, Wettstein, Sitzb. Akad. Wiss. Wien, 143:18, 1934.

Smith and Taylor (1950b:30) recorded this species from Sonora, Sinaloa, Nayarit, Colima, and Guerrero; these records indicate that the species probably is distributed along the Pacific coast of México southward from southern Sonora. It unquestionably occurs on the coast of Michoacán. Natives of the coastal lowlands tell of another "tortuga de la tierra" besides *Geoemyda rubida*. In the collections of the Museum of Natural History of the University of Illinois is a specimen of *Geoemyda pulcherrima* from Mexcala in the Balsas Basin in northern Guerrero. On the basis of this specimen it is highly probable that the species also inhabits the Balsas-Tepalcatepec Basin in Michoacán.

PSEUDEMYS SCRIPTA ORNATA (GRAY)

Emys ornata Gray, Synopsis reptilium, p. 30, 1831.—Mazatlán, Sinaloa, México.

Pseudemys scripta ornata, Carr, Herpetologica, 1:135, December 30, 1938.

The systematics and distribution of *Pseudemys scripta* in México and Central America are poorly understood. Smith and Taylor (1950b:32) recorded this turtle from the Pacific lowlands of Sinaloa, Jalisco, Oaxaca, and Chiapas. This species is represented by vicarious populations throughout the Atlantic lowlands of México, northwestern México, over much of the United States, and also in Baja California. Along the Pacific coast of México the species seems to be extremely rare, or, at least, only locally abundant. Since the species has such a wide distribution, and since it occurs on the Pacific lowlands both to the north and to the south of Michoacán, it is reasonable to expect its presence on the coast of Michoacán. Inquiries among the natives living in the Balsas-Tepalcatepec Basin produced only negative evidence about the occurrence of *Pseudemys* in the Río Tepalcatepec and Río Balsas. I suspect that the best place to search for these turtles on the coast of Michoacán is in the numerous fresh-water lagoons on the coastal plain.

CAIMAN CROCODILUS FUSCUS (COPE)

Perosuchus fuscus Cope, Proc. Acad. Nat. Sci. Philadelphia, 20:203, November 9, 1868.—Río Magdalena, Columbia.

Caiman crocodilus fuscus, Mertens, Senckenbergiana, 26:275, December 22, 1943.

Gadow (1930:50) reported that *Caiman sclerops* (= *Caiman crocodilus fuscus*) inhabited the "tierra caliente" in Michoacán. Smith and Taylor (1950b:212) accepted Gadow's record for the State, although otherwise the species is unknown north of Oaxaca. Peters (1954:10) refuted Gadow's record on the basis that Gadow's collections contained no specimens of *Caiman.* The local name "caiman" refers to both *Crocodylus* and to *Caiman,* for, in general, the natives do not distinguish between the two. "Caimanes" are reported from along the coast of Michoacán, where the name presumably refers to *Crocodylus acutus acutus,* and in the Balsas-Tepalcatepec Basin (Gadow, 1930:50; Webber, 1946:267). I have seen no specimens of either *Crocodylus* or *Caiman* from the Balsas Basin. If crocodilians do occur in the basin, they probably are *Crocodylus acutus acutus.* There is no basis, whatsoever, for including Michoacán in the range of *Caiman crocodilus fuscus.*

BIPES CANALICULATUS BONNATERRE

Bipes canaliculatus Bonnaterre, Encyclopédie méthodique, Erpétologie, p. 68, 1789.—México. Type locality restricted to Mexcala, Guerrero, México, by Smith and Taylor (1950b:39).

Dugès (1896:480) reported this species from Morelia, Michoacán. Smith and Taylor (1950b:39), who recorded the species from three localities in the Balsas Basin in Guerrero, rejected Dugès' record. I, too, am unwilling to accept Dugès' record. Nevertheless, the species probably occurs throughout much of the Balsas Basin. This idea is strengthened by comments made by Storm (1939:342): "The last hard drop, that afternoon, was down the great Cerro de los Cajones [southwest of

Tacámbaro], and here in the upper forest we came upon... a lizard with front legs and none behind ... the animal with hands and no feet that señor Smith [Hobart M. Smith] was seeking!... They're named *Bipes caniculatus* (*sic.*)."

COLEONYX ELEGANS NEMORALIS KLAUBER

Coleonyx elegans nemoralis Klauber, Trans. San Diego Soc. Nat. Hist., 10:195, March 9, 1949.—Paso del Río, Colima, México.

Klauber (1945:199) and Smith and Taylor (1950b:43) reported this lizard from the coastal lowlands of Colima and Guerrero. Davis and Smith (1953:101) reported it from 8 kilometers northeast of Temilpa, Morelos, in the upper Balsas Basin. Specimens of this lizard have been collected infrequently; the few locality records and limited ecological data indicate that it inhabits dense scrub forest and tropical semi-deciduous forest. *Coleonyx elegans nemoralis* is to be expected on the coastal lowlands, the seaward foothills of the Sierra de Coalcomán, and on the lower slopes of the Cordillera Volcánica along the northern edge of the Tepalcatepec Valley.

PHRYNOSOMA ORBICULARE ORBICULARE (LINNAEUS)

Lacerta obricularis Linnaeus, Systema naturae, ed. 12, 1:1062, 1789.—México (by inference). Type locality restricted to México, Districto Federal, by Smith and Taylor (1950b:97).

Phrynosoma orbiculare orbiculare, Smith, Trans. Kansas Acad. Sci., 37:290, 1934.

Gadow (1905:213) inferred that *Phrynosoma orbiculare* occurred at elevations of more than 3000 feet in Michoacán. There are no specimens of this species known from Gadow's collections made in Michoacán. Smith and Taylor (1950b:98) apparently accepted Gadow's statement and recorded the species from Michoacán: "above 3000 feet (Jorullo?)." Reeve (1952:940) somehow misconstrued this statement to read "Jorullo, above Zumpango (Smith and Taylor, 1950b)." Reeve did not indicate on his map (1952:939) that the species occurred in Michoacán. In the most recent review of the species (Horowitz, 1955), no localities are given in Michoacán. Since *Phrynosoma orbiculare* is known from central Jalisco, Guanajuato, Queretaro, and México, its presence at least in northeastern Michoacán is to be expected, although at the present time there are no specimens known from the state.

EUMECES BREVIROSTRIS (GÜNTHER)

Mabouia brevirostris Günther, Proc. Zool. Soc. London, p. 316, August, 1860.—Oaxaca. Type locality restricted to Oaxaca, Oaxaca, México, by Smith and Taylor (1950b:168).

Eumeces brevirostris, Bocourt, Mission scientifique au Mexique et dans l'Amerique Céntrale. Reptiles, livr. 6, p. 439, 1879.

Smith and Taylor (1950b:168) Listed this species: "*Michoacán*: No specific record." I am unaware of any specimen of this skink from the state. As presently recognized, this species contains two subspecies. One of these occurs in the mountains of Oaxaca northward into central Veracruz; the other, *Eumeces brevirostris bilineatus*, occurs in Durango southward to Jalisco, where it inhabits the Sierra Madre Occidental. Possibly the species occurs in the Sierra de los Tarascos in Michoacán.

EUMECES CALLICEPHALUS BOCOURT

Eumeces callacephalus Bocourt, Mission scientifique au Mexique et dans l'Amerique Céntrale. Reptiles, livr. 6, p. 431, 1879.— Guanajuato, Guanajuato, México.

Dugés (1896) in a paper in which he listed several species of *Eumeces* in México, reported *Eumeces callicephalus* from Michoacán, but he gave no specific locality within the state. Michoacán was included in the range of the species by Taylor (1936:298) and by Smith and Taylor (1950b:164). The species definitely is known from southeastern Arizona southward to Guanajuato. It may occur in Michoacán, but, since there are three rather widespread species of *Eumeces* inhabiting the Mexican Plateau and associated mountain ranges in the northern and northeastern part of Michoacán, interspecific competition might be a reason for the absence of *Eumeces callicephalus* there.

LEPTODEIRA SEPTENTRIONALIS POLYSTICTA GÜNTHER

Leptodeira polysticta Günther, Biologia Centrali-Americana, Reptilia, p. 172, May, 1895.—Belice, British Honduras.

Leptodeira septentrionalis polysticta, Duellman, Bull. Amer. Mus. Nat. Hist., 114:72, February 24, 1958.

Although this species occurs from sea level to elevations of about 2000 meters from Nayarit southward into Central America, no specimens are known from Michoacán. Smith and Taylor (1945:87) listed the species as occurring in Michoacán, but they had no record on which to base this report. Probably, the species occurs on the coastal lowlands and seaward slopes of the Sierra de Coalcomán.

TROPIDODIPSAS FASCIATA GUERREROENSIS TAYLOR

Tropidodipsas guerreroensis Taylor, Univ. Kansas Sci. Bull., 26:470; November 27, 1940.—Buena Vista, Guerrero, México.

Tropidodipsas fasciata guerreroensis, Alvarez del Toro and Smith, Herpetologica, 12:16, March 6, 1956.

Dugès (1896:480) reported a snake, questionably of this species, from Uruapan, Michoacán. Taylor (1940c) suggested that on geographic grounds Dugès' record might refer to *T. f. guerreroensis*, which is known definitely only from the type locality. *Tropidodipsas occidentala* is known from Comala, Colima, and Coalcomán, Michoacán. On zoogeopraphic grounds that species might be found at Uruapan. Since the specimen apparently no longer is extant, the identification cannot be ascertained.

MICRURUS FITZINGERI FITZINGERI (JAN)

Elaps fitzingeri Jan, Rev. Mag. Zool., p. 521, 1858.—México. Type locality restricted to Guanajuato, Guanajuato, México, by Smith and Taylor (1950a:330).

Micrurus fitzingeri fitzingeri, Brown and Smith, Proc. Biol. Soc. Washington, 55:63, June 25, 1942.

Smith and Taylor (1945:174) recorded the species from Zamora, Michoacán. Hobart M. Smith (*in litt.*) stated that this record was based on a report of *Elaps fulvius* from Zamora by Dugès (1896:482). Smith guessed that the report was based on a specimen of *Micrurus fitzingeri*. The specimen has not been seen. Although the species is known from Guanajuato and México, until a specimen is available from Michoacán, the species should not be considered part of the herpetofauna of Michoacán.

GAZETTEER

The localities in Michoacán here listed are those from which specimens were examined as well as other localities mentioned in the text. The localities are arranged alphabetically according to the most definitive word or words in the total name. For example, Lago de Chapala is listed as "Chapala (Lago de)" and Cerro de Tancítaro is listed as "Tancítaro (Cerro de)." Insofar as has been possible, the following information is given for each locality: geographical co-ordinates to the nearest minute of north latitude and west longitude, elevation in meters above mean sea level, a description of its geographical location, type of dominant vegetation, and in some cases comments concerning collecting sites in the vicinity. Distances are in kilometers; all are map (air line) distances, unless otherwise indicated. Many localities visited on mule trips are given as being a certain number of "mule hours" in a general direction from another town or village. In order to reach most of these localities today, one would have to go by mule, and this is the way the muleteers determine their distances. Some of the elevations are taken from maps, but most of them were obtained from one or more readings of altimeters that we carried in the field. The terms used for describing the vegetation are those defined in the section of the natural landscape.

My primary cartographic sources have been: the provisional edition of maps published by the American Geographic Society (Colima, Guadalajara, México, and San Luis Potosí sheets published between 1933 and 1940), scale 1:1,000,000; the preliminary sheets (Colima, Guadalajara, Guanajuato, and México) published in 1949 with a scale of 1:500,000 of the Carta Geográfica de la República Méxicana (Dirección de Geografía y Meterología, Secretaria de Agricultura y Ganadería); and the Carta de Cuenca Tepalcatepec (Scale 1:250,000) prepared in 1958 by the Comisión del Tepalcatepec, Secretaria de Recursos Hidráulicos. I have visited most of the 181 localities and have gathered data pertaining to vegetation, altitude, and location. I think, nevertheless, that the accuracy of some of the locations and elevations as given in the gazetteer is questionable. This situation can be rectified only by detailed geographic studies.

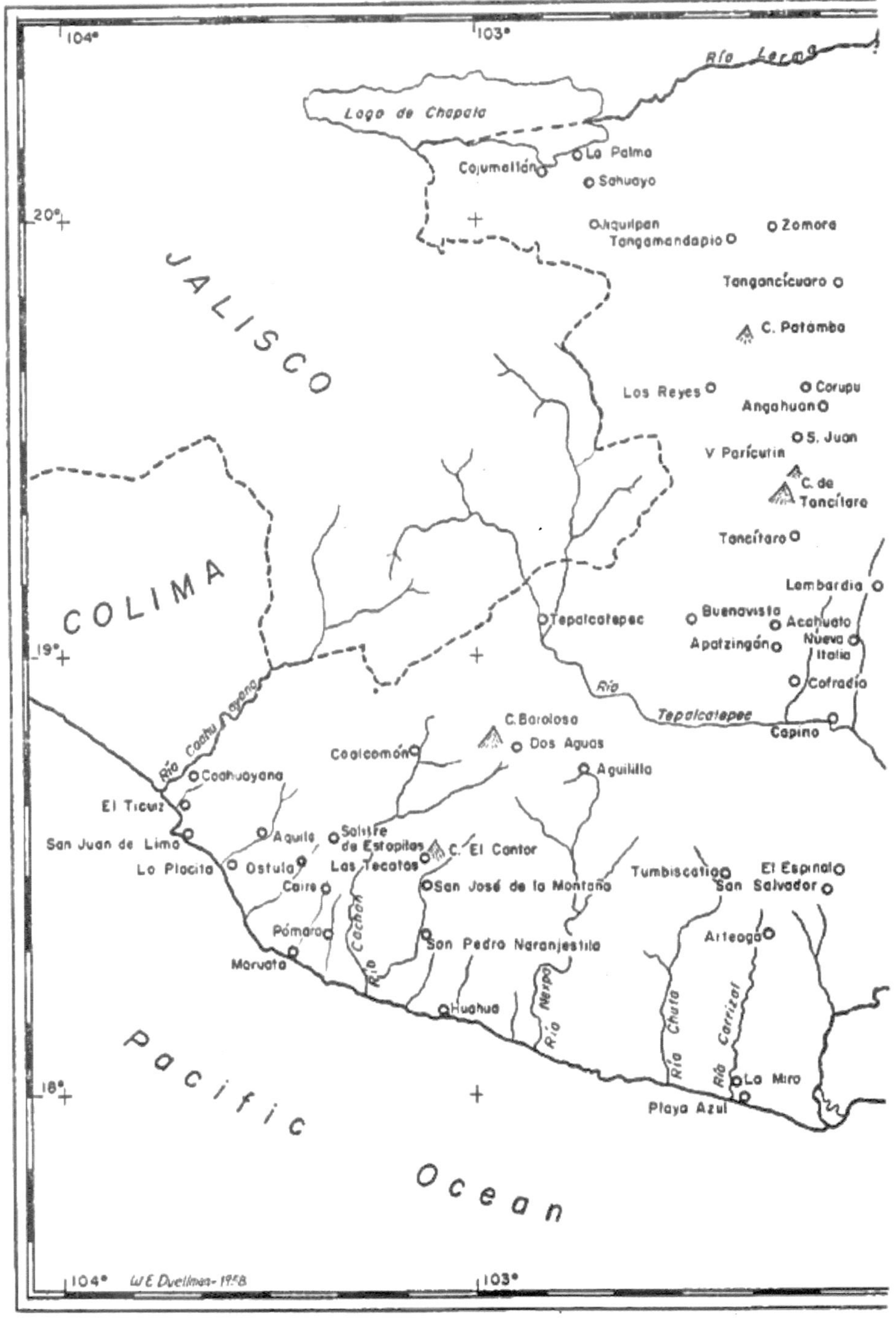

Fig. 11. Map of Michoacán showing important localities mentioned in text.

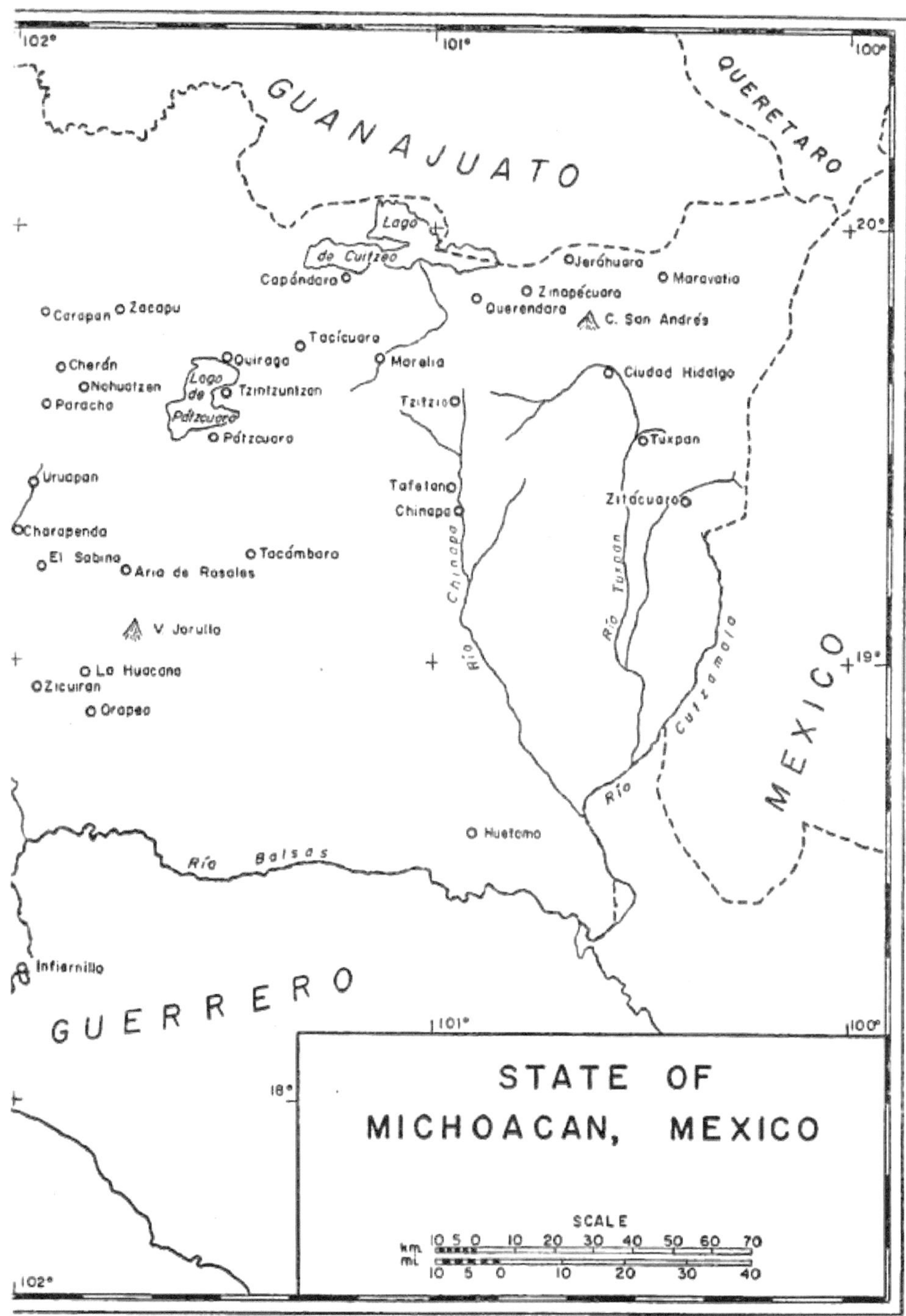

Localities not on this map can be located by directions given in the gazetteer.

Most of the important towns, villages, rivers, and high mountains are shown on the accompanying map (Fig. 11). Places not shown on this map can be located from directions given in the gazetteer.

Acahuata.—Lat. 19° 10', long. 102° 21', elev. 1040 m. A village north of Apatzingán and on the southern slope of Cerro de Tancítaro; transition between arid tropical scrub forest and pine-oak forest; tropical semi-deciduous forest in barrancas.

Agua Cerca.—Lat. 19° 06', long. 101° 45', elev. 1550 m. A ranch south-southwest of Ario de Rosales on the road to La Huacana; pine-oak forest.

Aguililla.—Lat. 18° 45', long. 102° 47', elev. 860 m.; a town in a low valley in the Sierra de Coalcomán; arid tropical scrub forest.

Álamo (El).—Lat. 19° 42', long. 100° 55', elev. 2300 m. A ranch 5 kilometers by road east of El Temazcal; pine-oak forest.

Angahuan.—Lat. 19° 33', long. 102° 14', elev. 2440 m. A Tarascan village about 27 kilometers northwest of Uruapan; pine forest. Much of the land is still covered by a deep layer of ashes from the nearby Volcán Parícutin.

Apatzingán.—Lat. 19° 06', long. 102° 22', elev. 335 m. The largest town in the Tepalcatepec Valley; arid tropical scrub forest.

Apiza (Boca de).—Lat. 18° 42', long. 103° 44', sea level. The name of the mouth of the Río Coahuayana; sandy beach and coco palms.

Apo.—Lat. 19° 25', long. 102° 25', elev. 2160 m. A village on the western slope of Cerro de Tancítaro; pine-oak forest.

Aquila.—Lat. 18° 32', long. 103° 30', elev. 150 m. A small village on the Río Aquila in the seaward foothills of the Sierra de Coalcomán; tropical semi-deciduous forest.

Araparicuaro.—Lat. 19° 22', long. 102° 12', elev. 1525 m. A village 19 kilometers west-southwest of Uruapan on the trail to Tancítaro; pine-oak forest.

Araro.—Lat. 19° 54', long. 100° 50', elev. 1830 m. A small village at the eastern end of the Lago de Cuitzeo lakebed; mesquite-grassland.

Ario de Rosales.—Lat. 19° 12', long. 101° 42', elev. 1980 m. A town in the Cordillera Volcánica on the road from Pátzcuaro to La Huacana; mixed hardwoods and pine forest.

Arteaga (formerly Carrizal).—Lat. 18° 28', long. 102° 25', elev. 850 m. A town in the eastern part of the Sierra de Coalcomán; transition between arid tropical scrub forest and oak forest.

Atzimba.—Lat. 19° 39', long. 100° 47', elev. 2900 m. A national park in the Cordillera Volcánica, located between Ciudad Hidalgo and Morelia, 32 kilometers by road west-southwest of Ciudad Hidalgo; mixed pine and fir forest.

Axolotl (Rancho).—Lat. 19° 47′, long. 100° 38′, elev. 2900 m. A settlement on the western slopes of Cerro San Andrés; pine, oak, and fir forest.

Balsas (Río).—A large river having its headwaters in Tlaxcala, Puebla, and northwestern Oaxaca, flowing westward through an arid valley to the Pacific Ocean, and in its lower part forming the boundary between Michoacán and Guerrero.

Barolosa (Cerro de).—Lat. 18° 52′, long. 102° 57′, elev. 2900-3050 m. Presumably the highest mountain in the Sierra de Coalcomán and located about 13 hours by mule east-northeast of Coalcomán; open pine-oak-fir forest and alder thickets.

Barolosa (Rancho).—Lat. 18° 50′, long. 103° 00′, elev. 2320 m. A small ranch on the west-northwestern slope of Cerro de Barolosa, about 11 hours by mule east-northeast of Coalcomán; open pine-oak forest.

Barranca Seca.—Lat. 19° 32′, long. 102° 15′, elev. 2100 m. A small village about 7 kilometers northwest of San Juan de Parangaricutiro; pine forest.

Bejuco (Barranca de).—Lat. 18° 07′, long. 102° 48′, elev. 90 m. A barranca in the lower slopes of the Sierra de Coalcomán just west of the lower reaches of the Río Nexpa; tropical semi-deciduous forest.

Buenavista (Tomatlán).—Lat. 19° 17′, long. 102° 36′, elev. 425 m. A village on the Río Masiaco in the Tepalcatepec Valley, 33 kilometers by road west-northwest of Apatzingán; open arid tropical scrub forest.

Buena Vista.—Lat. 18° 40′, long. 102° 09′, elev. 600 m. A ranch on the northeastern slopes of the Sierra de Coalcomán; arid tropical scrub forest.

Cachán (Río).—Lat. 18° 14′, long. 103° 14′. A river formed by the affluence of the Río Coalcomán and the Río San José and flowing into the Pacific Ocean at a point indicated by the co-ordinates given above. Sometimes the name is applied to the lower part of the river as used here; other times the name is used for the entire length of the Río Coalcomán.

Camichines.—Lat. 18° 47′, long. 103° 05′, elev. 1070 m. A ranch about 5 kilometers east-northeast of Coalcomán; transition between arid tropical scrub forest and oak forest.

Camécuaro (Lago de).—Lat. 19° 55′, long. 102° 13′, elev. 1615 m. A small lake (depth to about 10 m.) drained by the Río Duero and located one kilometer north-northwest of Tangancícuaro; mesquite-grassland and some cypress and oak around the lake.

Cancita (Río).—A tributary of the Río Tepalcatepec flowing southward from the southeastern slope of Cerro de Tancítaro.

Cantiles (Los).—Lat. 19° 43′, long. 100° 55′, elev. 2160 m. A ranch 33 kilometers by road east of Morelia; pine forest.

Capácuaro.—Lat. 19° 33′, long. 102° 02′, elev. 2070 m. A Tarascan village 18 kilometers by road north of Uruapan; pine forest.

Capirio.—Lat. 18° 52′, long. 102° 08′, elev. 180 m. A village on the Río Tepalcatepec, 22 kilometers by road south of Nueva Italia; open arid tropical scrub forest and some gallery forest along the river.

Carapan.—Lat. 19° 52′, long. 102° 02′, elev. 2070 m. A village on the northern edge of the Sierra de los Tarascos, 32 kilometers by road west of Zacapu; pine-oak forest at village and to the south; mesquite-grassland immediately to the north.

Cerrito (El).—Lat. 18° 45′, long. 103° 40′, elev. 15 m. A ranch about 3 kilometers northeast of Coahuayana; tropical semi-deciduous forest.

Chapala (Lago de).—A large lake on the Mexican Plateau at an elevation of 1525 m., partly in the state of Jalisco. It is drained by the Río Grande de Santiago, which flows northward and then westward into the Pacific Ocean. Immediately to the east of the lake are remnants of once extensive marshes.

Charapendo.—Lat. 19° 15′, long. 102° 04′, elev. 975 m. A village 24 kilometers by road south of Uruapan near the upper limit of the arid tropical scrub forest in the Tepalcatepec Valley.

Cherán.—Lat. 19° 42′, long. 101° 57′, elev. 2350 m. A Tarascan village 27 kilometers by road south-southeast of Carapan; pine forest.

Chichihuas.—Lat. 18° 47′, long. 103° 12′, elev. 1200 m. A ranch about 6 kilometers west-southwest of Coalcomán; scrub oak forest.

Chinapa.—Lat. 19° 22′, long. 100° 51′, elev. 930 m. A small village on the Río Chinapa, 43 kilometers south of El Temzcal on the road to Huetamo; arid tropical scrub forest.

Chupio.—Lat. 19° 10′, long. 101° 27′, elev. 1080 m. A village 12 kilometers by road south of Tacámbaro; transition between arid tropical scrub forest and oak forest.

Churumuco.—Lat. 18° 37′, long. 101° 38′, elev. 210 m. A small town in the Balsas Valley; arid tropical scrub forest.

Ciénega (La).—Lat. 18° 28′, long. 103° 18′, elev. 900 m. A ranch about 3 hours by mule north of Coire; tropical semi-deciduous forest.

Coahuayana.—Lat. 18° 44′, long. 103° 31′, elev. 15 m. A village on the coastal plain near the mouth of the Río Coahuayana; arid tropical scrub forest and tropical semi-deciduous forest.

Coalcomán.—Lat. 18° 47′, long. 103° 08′, elev. 945 m. The largest town in the Sierra de Coalcomán and situated in a valley about 12 by 6 kilometers; arid tropical scrub forest on valley floor; oaks and some tropical semi-deciduous forest on surrounding slopes.

Coalcomán (Río).—A river having its headwaters northeast of the town of Coalcomán and flowing southward to join with the Río San José to form the Río Cachán.

Coalcomán (Sierra de).—A highland mass outlined by the Río Coahuayana and its tributaries on the west, the Río Tepalcatepec on the north, and the Río Balsas on the east, and the Pacific Ocean on the south. The axis of the sierra extends for about 200 kilometers in a west-northwest to east-southeast direction; the mountains are nearly 80 kilometers in breadth; the highest parts of the range are about 3000 meters above sea level.

Cofradía.—Lat. 18° 56′, long. 102° 17′, elev. 215 m. A ranch about 17 kilometers southeast of Apatzingán; arid tropical scrub forest.

Coire.—Lat. 18° 26′, long. 103° 22′, elev. 300 m. A village on the seaward foothills of the Sierra de Coalcomán on the Río Coire; tropical semi-deciduous forest.

Cojumatlán.—Lat. 20° 07′, long. 102° 51′, elev. 1530 m. A village on the southeastern shore of Lago de Chapala; mesquite-grassland.

Colola (Río).—A small river emptying into the Pacific Ocean between Maruata and Punto San Telmo.

Conejos (Los).—Lat. 19° 22′ long. 102° 07′, elev. 1850 m. A village 6 kilometers west-southwest of Uruapan, and sometimes known as Nuevo San Juan; pine-oak forest.

Copándaro.—Lat. 19° 54′, long. 101° 12′, elev. 1800 m. A village on the south edge of the Lago de Cuitzeo lakebed; mesquite-grassland.

Copuyo (Capuyo or Copullo).—Lat. 18° 28′, long. 100° 56′, elev. 1200 m. A small village about 5 kilometers by road west of Paso Ancho; transition between arid tropical scrub forest and oak forest.

Cordillera Volcánica.—A mountain range along the southern edge of the Mexican Plateau, roughly along the nineteenth parallel, and made up of many volcanos; the range extends from Volcán de Colima on the west to Cofre de Perote and Orizaba in Veracruz; several of the volcanos reach elevations of more than 4000 meters.

Corralito (El).—Lat. 18° 52′, long. 102° 38′, elev. 270 m. A small village in the Tepalcatepec Valley, about 30 kilometers southwest of Apatzingán; arid tropical scrub forest.

Corupu (Corupo).—Lat. 19° 28′, long. 102° 19′, elev. 2450 m. A village

29 kilometers northwest of Uruapan; pine forest.

Cuatro Caminos.—Lat. 19° 00′, long. 102° 05′ elev. 335 m. A village 4 kilometers south of Nueva Italia; arid tropical scrub forest.

Cuilala (Playa).—Lat. 18° 10′, long. 103° 06′, sea level. A sandy beach on the Pacific Ocean just east of La Higuerita.

Cuitzeo.—Lat. 19° 58′, long. 101° 09′, 1800 m. A village on the north shore of the Lago de Cuitzeo lakebed; mesquite-grassland.

Cuitzeo (Lago de).—A large lakebed on the Mexican Plateau at an elevation of 1800 m. In dry years there is little water in the lake, and most of the lakebed is dry; in very wet years the entire lakebed is flooded. The Río de Morelia flows into the lake, which has no outlet; surrounding vegetation is mesquite-grassland.

Cuseño Station.—Lat. 19° 30′, long. 102° 16′, elev. 2200 m. A field station of the American Geological Society established in 1945 and demolished in 1953; located at the northern edge of the lava flow at Volcán Parícutin; remnants of pine forest.

Diezmo (El).—Lat. 18° 26′, long. 103° 19′, elev. 850 m. A ranch about 8 kilometers north of Coire; tropical semi-deciduous forest.

Dos Aguas.—Lat. 18° 45′, long. 102° 55′, elev. 2100 m. A lumber camp on the eastern slope of Cerro de Barolosa, located about 22 kilometers west-northwest of Aguililla; pine-oak forest and some fir forest in sheltered ravines.

Duero (Río).—A small river having its headwaters near Tangancícuaro and flowing northwestward into the Río Lerma; source of irrigation water for surrounding agricultural area.

Emiliano Zapata.—Lat. 18° 59′, long. 102° 39′ elev. 1600 m. A town 10 kilometers east of Jiquilpan; mesquite-grassland and irrigated fields.

Erongaricuaro.—Lat. 19° 35′, long. 101° 43′, elev. 2150 m. A Tarascan village on the western shore of Lago de Pátzcuaro; pine forest.

Espinal (El).—Lat. 18° 27′, long. 102° 07′, elev. 500 m. A ranch in the northern foothills of the Sierra de Coalcomán, 9 kilometers by road north-northeast of San Salvador; arid tropical scrub forest.

Estopilas (Salitre de).—Lat. 18° 30′, long. 103° 23′, elev. 130 m. A small village about 10 kilometers east of Ostula; tropical semi-deciduous forest and arid tropical scrub forest.

Garnica (Cerro).—Lat. 19° 43′, long. 100° 48′, elev. 3000 m. A mountain about 8 kilometers north of Pino Gordo; pine-oak-fir forest.

Garnica (Puerto de).—Lat. 19° 42′, long. 100° 51′, elev. 2840 m. A mountain pass 46 kilometers by road west of Ciudad Hidalgo; pine and fir forest.

Gregorio (San).—Lat. 19° 25′, long. 101° 24′, elev. 2200 m. A ranch about 16 kilometers southeast of Pátzcuaro; pine forest.

Guayabo.—Lat. 18° 45′, long. 102° 15′, elev. 760 m. A village in the Sierra de Coalcomán about 32 kilometers north-northeast of Arteaga; upper limits of arid tropical scrub forest.

Herradero (Barranca de).—Lat. 18° 17′, long. 103° 08′, elev. 200-250 m. A barranca south of San Pedro Naranjestila in the Sierra de Coalcomán; tropical semi-deciduous forest.

Hidalgo (Ciudad).—Lat. 19° 32′, long. 100° 34′, elev. 2100 m. A town in the valley of the Río Tuxpan; mesquite-grassland and pine-oak forest.

Higuerita (La).—Lat. 18° 12′, long. 103° 06′, sea level. A place name on the Pacific coast; sandy beach and arid tropical scrub forest.

Higuertas (Las).—Lat. 18° 39′, long. 103° 17′, elev. 1600 m. A ranch about 7 hours by mule southwest of Coalcomán; pine-oak forest.

Hondo (Puerto).—Lat. 19° 25′, long. 100° 13′, elev. 2750 m. A pass in the mountains, 14 kilometers by road east of Zitácuaro (just west of Macho de Agua); pine, oak, and fir forest.

Huancana (La).—Lat. 18° 58′, long. 101° 50′, elev. 550 m. A village in the Balsas Basin; arid tropical scrub forest.

Huahua (La).—Lat. 18° 12′, long. 103° 00′, sea level. A small village on the Pacific coast; arid tropical scrub forest and gallery forest along the Arroyo de Huahua.

Huetamo.—Lat. 18° 38′, long. 100° 53′, elev. 300 m. A town in the Balsas Valley; arid tropical scrub forest.

Huingo.—Lat. 19° 55′, long. 100° 50′, elev. 1800 m. A village on the eastern edge of the Lago de Cuitzeo lakebed; mesquite-grassland.

Jacona.—Lat. 19° 57′, long. 102° 18′, elev. 1600 m. A small town, 4.3 kilometers by road southwest of Zamora; mesquite-grassland.

Jaramillo.—Lat. 19° 20′, long. 102° 02′, elev. 1500 m. A ranch 9 kilometers by road south of Uruapan; pine-oak forest.

Jazmin.—Lat. 18° 52′, long. 101° 58′, elev. 275 m. A village in the Tepalcatepec Valley, 32 kilometers by road southeast of Cuatro Caminos; open arid tropical scrub forest.

Jeráhuaro.—Lat. 19° 52′, long. 100° 35′, elev. 2600 m. A town in the northern part of the state and located east of Lago de Cuitzeo; pine-oak forest.

Jiquilpan.—Lat. 19° 59′, long. 102° 43′, elev. 1570 m. A town just southeast of Lago de Chapala; mesquite-grassland.

Jorullo (Volcán).—Lat 19° 00′, long. 101° 45′, elev. 1300 m. (crest). A cinder and lava cone rising from the foothills of the Cordillera Volcánica; arid tropical scrub forest on lower slopes and pine-oak forest on top.

Jungapeo.—Lat. 19° 26′, long. 100° 29′, elev. 1430 m. A village in the valley of the Río Tuxpan, about 13 kilometers south of Tuxpan on the southern slopes of the Mexican Plateau; tropical semi-deciduous forest and pine-oak forest.

Lengua de Vaca (Puerto de).—Lat. 19° 26′, long. 100° 13′, elev. 2900 m. A pass in the mountains at the Michoacán-Mexico border through which passes the Mexico City-Morelia highway; pine and fir forest.

Lerma (Río).—A river originating in the state of México and flowing westward, and forming the northern boundary of the state of Michoacán, to Lago de Chapala.

Lima (San Juan de).—Lat. 18° 29′, long. 102° 42′, sea level. A ranch on the Pacific coast; arid tropical scrub forest and tropical semi-deciduous forest.

Lima (Punta San Juan de).—Lat. 18° 38′, long. 102° 43′, sea level. A rocky promontory jutting into the Pacific Ocean just southwest of San Juan de Lima; arid tropical scrub forest.

Limoncito.—Lat. 18° 45′, long. 102° 43′, elev. 730 m. A ranch 10 kilometers north of Aguililla; arid tropical scrub forest; tropical semi-deciduous gallery forest along the nearby Río Tepecuate.

Lombardia.—Lat. 19° 08′, long. 102° 02′, elev. 640 m. A town in the Tepalcatepec Valley, 38 kilometers by road south of Uruapan; arid tropical scrub forest.

Lleguas (Acuaro de las).—Lat. 18° 48′, long. 102° 52′, elev. 2320 m. A place name for a stream and meadow (Llano de la Llegua) surrounded by pine-oak forest, located about 10 hours by mule east of Coalcomán.

Macho de Agua.—Lat. 19° 25′, long. 100° 15′, elev. 2850 m. A ranch just west of Puerto de Lengua de Vaca and 16 kilometers by road east of Zitácuaro; mixed oak, pine, and fir forest.

Maquili.—Lat. 18° 36′, long. 103° 32′, elev. 120 m. A village on the Río Aquila about 3 kilometers south-southwest of Aguila; tropical semi-deciduous forest.

Maravatio.—Lat. 19° 53′, long. 100° 27′, elev. 2010 m. A town in the Río Lerma Valley; irrigated fields on flats and pine-oak forest on slopes.

Marquez (Río).—A tributary to the Río Tepalcatepec, flowing through a deep gorge (Barranca del Marquez) between Lombardia and Nueva Italia. The stream originates from springs near Uruapan, where the stream is known as the Río Cupatitzio.

Maruata.—Lat. 18° 17′, long. 103° 20′, sea level. Place name for a Nine-teenth Century port of little importance near the mouth of the Río Coire; sandy beach, fresh-water lagoon, and arid tropical scrub forest.

Mexcala (Laguna).—Lat. 18° 29′, long. 103° 41′, sea level. A brackish lagoon surrounded by mangroves, located just southwest of El Ticuiz.

Mil Cumbres.—Lat. 19° 39′, long. 100° 47′, elev. 2800 m. A name for a look-out on the México-Morelia highway in Atzimba National Park, about 32 kilometers by road west-southwest of Ciudad Hidalgo; pine and fir forest.

Mira (La).—Lat. 18° 05′, long. 102° 20′, elev. 20 m. A small village about 5 kilometers north-northeast of Playa Azul; arid tropical scrub forest.

Morelia.—Lat. 19° 43′, long. 101° 10′, elev. 1900 m. Capital of and largest city in Michoacán; mesquite-grassland on flats and pine-oak forest on surrounding hills.

Morelia (Río de).—A small, intermittent stream originating in the mountains south of Morelia and emptying into Lago de Cuitzeo.

Motín del Oro.—Lat. 18° 14′, long. 103° 48′, sea level. A ranch on the Pacific coast; arid tropical scrub forest.

Motín (Río).—Lat. 18° 13′, long. 103° 48′ (mouth). A small river flowing from the Sierra de Coalcomán into the Pacific Ocean.

Nahuatzen (Nauhuatzin).—Lat. 19° 42′, long. 101° 50′, elev. 2450 m. A Tarascan village in the mountains west of Lago de Pátzcuaro; pine forest.

Nexpa (Río).—Lat. 18° 05′, long. 102° 47′ (mouth). A large river draining the central part of the Sierra de Coalcomán, originating near Aguililla, and flowing into the Pacific Ocean.

Nogueleras.—Lat. 18° 34′, long. 103° 17′, elev. 1600 m. A ranch about 10 hours by mule south-southwest of Coalcomán; oak forest.

Nueva Italia.—Lat. 19° 02′, long. 102° 07′, elev. 380 m. A town in the Tepalcatepec Valley, 59 kilometers by road south of Uruapan; arid tropical scrub forest.

Nuevo (Rancho).—Lat. 18° 26′, long. 102° 07′, elev. 520 m. A ranch 7 kilometers by road north-northeast of San Salvador in the northern foothills of the Sierra de Coalcomán; arid tropical scrub forest.

Ocorla.—Lat. 18° 38′, long. 103° 07′, elev. 885 m. A ranch about 6 hours by mule south-southeast of Coalcomán; scrubby oak forest.

Opopeo.—Lat. 19° 24′, long. 101° 37′, elev. 2800 m. A village 16 kilometers south of Pátzcuaro; pine and fir forest.

Orilla (La).—Lat. 18° 00′, long. 102° 12′, elev. 10 m. The site of a former hacienda of the same name near the mouth of the Río Balsas; arid tropical scrub forest.

Oropeo.—Lat. 18° 52′, long. 101° 48′, elev. 300 m. A village in the Tepalcatepec Valley about 13 kilometers south of La Huacana; arid tropical scrub forest.

Ostula.—Lat. 18° 30′, long. 103° 28′, elev. 120 m. A village in the seaward foothills of the Sierra de Coalcomán, located on the Río Ostula about 16 kilometers east-southeast of La Placita; arid tropical scrub forest and scattered tropical semi-deciduous forest.

Ozumatlán (Sierra de).—A range in the Cordillera Volcánica extending east-northeast from a point south of Morelia to Queréndaro and reaching elevations in excess of 2600 m.

Palma (La).—Lat. 20° 09′, long. 102° 46′, elev. 1525 m. A village on the southeastern shore of Lago de Chapala; lake-shore marshes and mesquite-grassland.

Paracho.—Lat. 19° 39′, long. 102° 02′, elev. 2375 m. A Tarascan village in the Cordillera Volcánica, located 35 kilometers by road north of Uruapan; pine forest.

Parangaricutiro (San Juan de).—Lat. 19° 30′, long. 102° 15′, elev. 2200 m. A former Tarascan village that was destroyed by the eruption of Volcán Parícutin; lava and volcanic ash amidst open pine forest.

Parícutin (Volcán).—Lat. 19° 30′, long. 102° 16′, elev. 2200 m. at base and 2700 m. at summit. A volcano born in February, 1943; it ceased to be active in December, 1951, and is located at the north-northeastern base of Cerro de Tancítaro; volcanic ash and lava amidst open pine forest.

Paso Ancho.—Lat. 19° 28′, long. 100° 52′, elev. 1100 m. A small village 30 kilometers south of El Temazcal on the road to Huetamo; arid tropical scrub forest.

Patamba (Sierra).—Lat. 19° 45′, long. 102° 21′, elev. 3700 m. at summit. A mountain, the summit of which is about 22 kilometers southwest of Tangancícuaro; pine forest from 2000 to 2600 m.; fir forest above 2600 m.

Pátzcuaro.—Lat. 19° 30′, long. 101° 36′, elev. 2200 m. A town near the southeastern shore of Lago de Pátzcuaro; pine forest.

Pátzcuaro (Lago de).—A large lake on the southwestern part of the Mexican Plateau at an elevation of 2165 m. It has no outlet. The lake is surrounded by mountains supporting pine and pine-oak forest. Along the southern and eastern shores of the lake are small marshes.

Peñas (Las).—Lat. 18° 03′, long. 102° 38′, sea level. A small village on the Pacific coast; arid tropical scrub forest.

Pichi (Estero).—Lat. 18° 01′, long. 102° 24′, sea level. A brackish lagoon surrounded by mangroves and coconut groves, located just east of Playa Azul.

Pino Gordo.—Lat. 19° 42′, long. 100° 45′, elev. 2600 m. A ranch 37 kilometers by road west of Ciudad Hidalgo; pine-oak forest.

Placita (La).—Lat. 18° 32′, long. 103° 37′, elev. 20 m. A village on the coastal lowlands, located on the Río Aquila; arid tropical scrub forest; tropical semi-deciduous forest along the river.

Playa (La).—Lat. 18° 57′, long. 102° 33′, elev. 800 m. A small village on the western edge of the lava flow of Volcán Jorullo; arid tropical scrub forest and some tropical semi-deciduous forest in ravines.

Playa Azul.—Lat. 18° 01′, long. 102° 25′, sea level. A village on the Pacific coast near the mouth of the Río Carrizal; arid tropical scrub forest; coconut plantations; mangrove-lined lagoons.

Pómaro.—Lat. 18° 18′, long. 103° 17′, elev. 300 m. An Indian village in the southern foothills of the Sierra de Coalcomán, located about 3 hours by mule north-northeast of Maruata; tropical semi-deciduous forest.

Pozos (Los).—Lat. 18° 30′, long. 103° 17′, elev. 300 m. A ranch located about 5 hours by mule north of Coire; tropical semi-deciduous forest.

Queréndaro.—Lat. 19° 48′, long. 100° 53′, elev. 1900 m. A town on the Mexican Plateau south of Lago de Cuitzeo; mesquite-grassland.

Quiroga.—Lat. 19° 42′, long. 101° 30′, elev. 2200 m. A Tarascan town on the north edge of Lago de Pátzcuaro; mesquite-grassland and pine-oak forest.

Reyes (Los).—Lat. 19° 35′, long. 102° 28′, elev. 1500 m. A town in western Michoacán, 50 kilometers south-southwest of Zamora; mesquite-grassland, oak and pine forest.

Sabino (El).—Lat. 19° 14′, long. 102° 03′, elev. 1050 m. A hacienda about 24 kilometers south of Uruapan; arid tropical scrub forest, many streams, rice fields.

Sahuayo.—Lat. 20° 05′, long. 102° 43′, elev. 1550 m. A town just south of the eastern end of Lago de Chapala; mesquite-grassland.

Salada (La).—Lat. 19° 07′, long. 102° 00′, elev. 580 m. A ranch southwest of Lombardia; arid tropical scrub forest.

Salto (Arroyo El).—Lat. 18° 45′, long. 103° 04′, elev. 1370 m. A valley of the Río Flores about 3 hours by mule east-southeast of Coalcomán; pine-oak forest.

San Andrés (Cerro).—Lat. 19° 48′, long. 100° 35′, elev. 3950 m. at summit. A mountain, the summit of which is about 16 kilometers north-northwest of Ciudad Hidalgo; oak forest to 2500 m. and pine and fir forest above 2500 m.

San José (de la Cumbre).—Lat. 19° 41′, long. 100° 50′, elev. 2750 m. A ranch 51 kilometers by road east of Morelia; pine and fir forest.

San José (de la Montaña).—Lat. 18° 25′, long. 103° 06′, elev. 750 m. A village sometimes called La Guitarra, located 14 hours by mule south-southeast of Coalcomán; tropical semi-deciduous forest.

San Pedro Naranjestila.—Lat. 18° 17′, long. 103° 06′, elev. 500 m. An Indian village in the southern foothills of the Sierra de Coalcomán; tropical semi-deciduous forest.

San Salvador.—Lat. 18° 25′, long. 102° 08′, elev. 700 m. A small village in the Sierra de Coalcomán, 37 kilometers by road northeast of Arteaga; arid tropical scrub forest.

San Telmo (Ojos de Agua de).—Lat. 18° 37′, long. 103° 42′, sea level. A small settlement at the base of Punto San Juan de Lima; tropical semi-deciduous forest and groves of oil palms.

San Telmo (Punta).—Lat. 18° 18′, long. 103° 29′, sea level. A rocky promontory jutting into the Pacific Ocean, on which there is a lighthouse (El Faro); arid tropical scrub forest.

Santa Ana.—Lat. 18° 27′, long. 102° 06′, elev. 600 m. A ranch about 4 kilometers by road northeast of San Salvador; arid tropical scrub forest.

Tacámbaro.—Lat. 19° 05′, long. 101° 22′, elev. 1820 m. A town in the Cordillera Volcánica; pine forest.

Tacícuaro.—Lat. 19° 38′, long. 101° 18′, elev. 2000 m. A village 21 kilometers east-southeast of Quiroga; mesquite-grassland and scrubby oak forest.

Tafetan.—Lat. 19° 43′, long. 100° 52′, elev. 1000 m. A village 40 kilometers by road south of El Temazcal; arid tropical scrub forest.

Tancítaro.—Lat. 19° 20′, long. 102° 22′, elev. 1850 m. A small town on the southern slope of Cerro de Tancítaro; pine-oak forest.

Tancítaro (Cerro de).—Lat. 19° 25′, long. 102° 18′, elev. 3870 m. at summit. An old volcano in the Cordillera Volcánica; the southern slope drops into the Tepalcatepec Valley; the summit is about 30 kilometers west of Uruapan; pine and oak forest on lower slopes replaced by pine or fir forest above.

Tangamandapio.—Lat. 19° 56′, long. 102° 25′, elev. 1700 m. A small town on the Mexican Plateau between Jiquilpan and Zamora; mesquite-grassland and irrigated fields.

Tangancícuaro.—Lat. 19° 52′, long. 102° 13′, elev. 1770 m. A town 12 kilometers by road southeast of Zamora; mesquite-grassland and irrigated fields.

Tarascos (Sierra de los).—A name applied to that part of the Cordillera Volcánica extending eastward from Cerro de Tancítaro and Sierra Patamba to Pátzcuaro.

Tarécuaro.—Lat. 19° 53′, long. 102° 29′, elev. 1700 m. A village on the Mexican Plateau, 26 kilometers southwest of Zamora; mesquite-grassland and pine-oak forest.

Tecatas (Las).—Lat. 18° 36′, long. 103° 17′, elev. 1950 m. A ranch located about 10 hours by mule south-southwest of Coalcomán; oak forest.

Temazcal (El).—Lat. 19° 40′, long. 100° 56′, elev. 2200 m. A road junction, 29 kilometers east of Morelia; here the road to Huetamo leads south from the Mexico City-Morelia highway; pine forest.

Tepalcatepec.—Lat. 19° 10′, long. 102° 50′, elev. 570 m. A village in the upper Tepalcatepec Valley; arid tropical scrub forest.

Tepalcatepec (Río).—A large river having its headwaters in southeastern Jalisco and flowing through a broad valley, which separates the Cordillera Volcánica from the Sierra de Coalcomán, to the Río Balsas.

Ticuiz (El).—Lat. 18° 40′, long. 103° 40′, elev. 10 m. A village on the coastal plain about 11 kilometers south of Coahuayana; arid tropical scrub forest and tropical semi-deciduous forest.

Tinguidín.—Lat. 19° 45′, long. 102° 28′, elev. 1800 m. A small town, 17 kilometers north of Los Reyes; pine-oak forest.

Tizupan (Río).—Lat. 18° 09′, long. 102° 55′ (mouth). A small river flowing southward from the Sierra de Coalcomán to the Pacific Ocean.

Tlalpujahua.—Lat. 19° 48′, long. 100° 10′, elev. 2600 m. A mining town in the northeastern part of the state; pine and fir forest.

Tumbiscatio.—Lat. 18° 32′, long. 102° 20′, elev. 900 m. A town in the Sierra de Coalcomán; arid tropical scrub forest.

Tupátaro.—Lat. 19° 53′, long. 100° 15′, elev. 2050 m. A village in the northeastern corner of the state, 13 kilometers northwest of Tlalpujahua; oak forest.

Tuxpan.—Lat. 19° 35′, long. 100° 27′, elev. 1850 m. A town in a basin nearly surrounded by mountains and near the headwaters of the Río Tuxpan, 19 kilometers by road east-southeast of Ciudad Hidalgo; arid mesquite-grassland and irrigated fields.

Tuxpan (Río).—A river draining the mountains in the eastern part of the state and flowing southward into the Río Balsas.

Tzararacua (Cascada).—Lat. 19° 18′, long. 102° 02′, 1430 m. A waterfalls of the Río Cupatitzio, 10.5 kilometers by road south of Uruapan; oak forest with scattered pines.

Tzintzuntzan.—Lat. 19° 38′, long. 101° 35′, elev. 2170 m. A village at the site of the seat of the ancient Tarascan empire on the eastern shore of Lago de Pátzcuaro; grasslands and marshes.

Tzitzio.—Lat. 19° 35′, long. 100° 55′, elev. 1630 m. A village 16 kilometers by road south of El Temazcal; pine-oak and arid tropical scrub forest.

Ucareo (Serranía de).—A part of the Cordillera Volcánica, including Cerro San Andrés.

Undameo.—Lat. 19° 34′, long. 101° 17′, elev. 2000 m. A village 20 kilometers west-southwest of Morelia; mesquite-grassland.

Uruapan.—Lat. 19° 25′, long. 102° 02′, elev. 1630 m. A large town on the southern slopes of the Cordillera Volcánica; pine-oak forest.

Zacapu.—Lat. 19° 48′, long. 101° 47′, elev. 2000 m. A town on the Mexican Plateau; mesquite-grassland.

Zamora.—Lat. 19° 59′, long. 102° 17′, elev. 1570 m. A large town on the Mexican Plateau; mesquite-grassland.

Zicuiran.—Lat. 18° 53′, long. 101° 55′, elev. 190 m. A small village 23 kilometers east-southeast of Cuatro Caminos; arid tropical scrub forest.

Zinapécuaro.—Lat. 19° 52′, long. 100° 49′, elev. 1900 m. A town near the southeastern end of Lago de Cuitzeo; mesquite-grassland and pine-oak forest.

Ziracuaretiro.—Lat. 19° 25′, long. 101° 52′, elev. 1230 m. A village 19 kilometers by road east of Uruapan; transition between pine-oak forest and arid tropical scrub forest.

Zirimícuaro.—Lat. 19° 24′, long. 101° 56′, elev. 1300 m. A hacienda 13 kilometers by road east of Uruapan; pine-oak forest and fields of sugar cane.

Zitácuaro.—Lat. 19° 25′, long. 100° 21′, elev. 2100 m. A town in the highlands of eastern Michoacán; pine-oak forest.

Zurumbeneo.—Lat. 19° 43′, long. 101° 02′, elev. 2100 m. A ranch 19 kilometers by road east of Morelia; scrubby oak forest.

SUMMARY

The preceding analysis of the amphibians and reptiles of the state of Michoacán shows that the herpetofauna is composed of 176 species and subspecies definitely recorded from the state, plus ten others that probably occur there. Ten species are reported for the first time from Michoacán: *Pseudoeurycea robertsi, Leptodactylus occidentalis, Microbatrachylus pygmaeus, Pternohyla fodiens, Hypopachus caprimimus, Phyllodactylus homolepidurus, Anolis dunni, Sceloporus bulleri, Sceloporus heterolepis,* and *Geagras redimitus.* Five species that have been reported previously from Michoacán are based on specimens having unreliable locality data or on misidentifications; therefore, the following species are not considered to be a part of the herpetofauna of Michoacán: *Caiman crocodilus fuscus, Urosaurus irregularis, Geophis nasalis, Tropidodipsas fasciata guerreroensis,* and *Micrurus fitzingeri fitzingeri.*

Systematic studies based at least in part on specimens from Michoacán have resulted in a redefinition of nine species and subspecies: *Bufo marmoreus, Bufo perplexus, Anolis nebulosus, Anolis nebuloides, Sceloporus bulleri, Sceloporus heterolepis, Sceloporus melanorhinus calligaster, Hypsiglena torquata torquata,* and *Hypsiglena torquata ochrorhyncha.*

Nine species that previously have been recognized as valid have been placed in synonymy. These are: *Bufo horribilis* Wiegmann, 1833, and *Bufo angustipes* Smith and Taylor, 1945, as synonyms of *Bufo marinus* (Linnaeus), 1758. *Microbatrachylus albolabris* Taylor, 1940, *Microbatrachylus minimus* Taylor, 1940, and *Microbatrachylus imitator* Taylor, 1942, as synonyms of *Microbatrachylus pygmaeus* (Taylor), 1937. *Phrynohyas corasterias* Shannon and Humphrey, 1957, as a synonym of *Phrynohyas inflata* (Taylor), 1944. *Hyla microeximia* Maslin, 1957, as a synonym of *Hyla eximia* Baird, 1854. *Hylella azteca* Taylor, 1943, as a synonym of *Hyla smaragdina* Taylor, 1940. *Loxocemus sumichrasti* Bocourt, 1876, as a synonym of *Loxocemus bicolor* Cope, 1861. *Eleutherodactylus vocalis* Taylor, 1940, is considered to be a subspecies of *Eleutherodactylus rugulosus.* The populations of *Thamnophis dorsalis* in the Tepalcatepec Valley are shown to be distinct from those inhabiting the highlands of the state; *Thamnophis dorsalis postremus* Smith, 1942, is revived for the population in the Tepalcatepec Valley.

Descriptions are given of the tadpoles of *Bufo occidentalis* and *Hyla bistincta.*

LITERATURE CITED

ANDERSON, J. D.

1960. Storeria storerioides in western Mexico. Herpetologica, 16:63-6, March 31.

BOGERT, C. M., and OLIVER, J. A.

1945. A preliminary analysis of the herpetofauna of Sonora. Bull. Amer. Mus. Nat. Hist., 83:297-426, March 30.

BOULENGER, G. A.

1885. Catalogue of the lizards in the British Museum, vol. II. London, xiii + 497 pp., November 15.

1894. Catalogue of the snakes in the British Museum, vol. II, London, xi + 382 pp., September 23.

CONANT, R.

1953. Three new water snakes of the genus Natrix from Mexico. Nat. Hist. Misc., 126:1-9, September 15.

CONTRERAS ARIAS, A.

1942. Mapa de las provincias climatologicas de la República Méxicana. Dir. Meteoro, Hidro. Inst. Geog. México, xxvii + 54 pp.

COPE, E. D.

1861. Contributions to the ophidiology of Lower California, Mexico, and Central America. Proc. Acad. Nat. Sci. Philadelphia, 13:292-306, December 28.

1887. Catalogue of batrachians and reptiles of Central America and Mexico. Bull. U. S. Natl. Mus., 32:1-98.

DAVIS, W. B.

1954. Three new anoles from Mexico. Herpetologica, 10:1-6, April 20.

DAVIS, W. B., and DIXON, J. R.

1957. Notes on Mexican snakes (Ophidia). Southwest. Nat., 2:19-27, January.

Davis, W. B., and Smith, H. M.

1953. Lizards and turtles of the Mexican state of Morelos. Herpetologica. 9:100-108, July 22.

Dixon, J. R.

1957. Geographic variation and distribution of the genus Tomodactylus in Mexico. Texas Jour. Sci., 9:379-409, December.

1960. Two new geckos, genus Phyllodactylus (Reptilia: Sauria), from Michoacán, Mexico. Southwest. Nat., 5:37-42, April 15.

Dowling, H. G.

1960. A taxonomic study of the ratsnakes, genus Elaphe Fitzinger. VII. The triaspis section, Zoologica, 45:53-80, August 15.

Duellman, W. E.

1954a. The salamander Plethodon richmondi in southwestern Ohio. Copeia, 1950 (1):40-45, February 19.

1954b. The amphibians and reptiles of Jorullo Volcano, Michoacan, Mexico. Occ. Pap. Mus. Zool. Univ. Michigan, 560:1-24, October 22.

1955. A new whiptail lizard, genus Cnemidorphorus, from Mexico. Occ. Pap. Mus. Zool. Univ. Michigan, 574:1-7, December 23.

1956a. The frogs of the hylid genus Phrynohyas Fitzinger, 1843. Misc. Publ. Mus. Zool. Univ. Michigan, 96:1-47, February 21.

1956b. A new snake of the genus Leptotyphlops from Michoacán, México. Copeia, 1956 (2):93-94, May 29.

1957a. Sexual dimorphism in the hylid frog Agalychnis dacnicolor Cope, and the status of Agalychnis alcorni Taylor. Herpetologica, 13:29-30, March 30.

1957b. Notes on snakes from the Mexican state of Sinaloa. Herpetologica, 13:237-240, October 31.

1958a. A monographic study of the colubrid snake genus Leptodeira. Bull. Amer. Mus. Nat. Hist., 114:1-152, February 24.

1958b. Comments on the type locality and geographic distribution of Urosaurus gadowi. Copeia, 1958 (1):48-49, February 21.

1958c. A preliminary analysis of the herpetofauna of Colima, Mexico. Occ. Pap. Mus. Zool. Univ. Michigan, 589:1-22, March 21.

1959. Two new snakes, genus Geophis, from Michoacan, Mexico. Occ. Pap. Mus. Zool. Univ. Michigan, 605:1-9, May 29.

1960a. A new subspecies of lizard, Cnemidophorus sacki, from Michoacán, México. Univ. Kansas Publ. Mus. Nat. Hist., 10:587-

598, May 2.

1960b. A taxonomic study of the Middle American snake Pituophis deppei, Univ. Kansas Publ. Mus. Nat. Hist., 10:599-610, May 2.

1960c. Variation, distribution, and ecology of the Mexican teiid lizard Cnemidophorus calidipes. Copeia, 1960(2):97-101, June 29.

1960d. A distributional study of the amphibians of the Isthmus of Tehuantepec, México. Univ. Kansas Publ. Mus. Nat. Hist., 13:19-72, August 16.

DUELLMAN, W. E., and DIXON, J. R.

1959. A new frog of the genus Tomodactylus from Michoacán, Mexico. Texas Jour. Sci., 11:78-82, March.

DUELLMAN, W. E., and DUELLMAN, A. S.

1959. Notes on the variation, distribution, and ecology of the iguanid lizard Enyaliosaurus clarki. Occ. Pap. Mus. Zool. Univ. Michigan, 598:1-10, February 16.

DUELLMAN, W. E., and WELLMAN, J.

1960. A systematic study of the lizards of the deppei group (Genus Cnemidophorus) in Mexico and Guatemala. Misc. Publ. Mus. Zool. Univ. Michigan, 111:1-80, February 10.

DUGÈS, A.

1884. Dos reptiles de Mexico. La Naturaleza, 6:359-362.

1891. Eumeces altamirani A. Dug. La Naturaleza, ser. 2, 1:485-486.

1896. Reptiles y batracios de los E. U. Mexicanos. La Naturaleza, ser. 2, 2:479-485.

DUMÉRIL, A. M., and BOCOURT, F.

1870-1909. Mission scientifique au Mexique et dans l'Amerique Centrale, Recherches zoologiques. Études sur les reptiles. Paris, xiv + 1012 pp.

FISCHER, J. G.

1882. Herpetologische Bemerkengen vorzugsweise über Stücke der Sammlung des Naturhistorischen Museum in Bremen. Abhand. Naturwiss. Ver. Bremen, 7:225-238.

1883. Beschreibungen neuer Reptilien. Akad. Gymnas. Hamburg, 1883:1-16.

FITCH, H. S., and MILSTEAD, W. W.

1961. An older name for Thamnophis cyrtopsis (Kennicott). Copeia, no. 1:112, March 17, 1961.

FUNKHOUSER, A.

1957. A review of the Neotropical tree-frogs of the genus Phyllomedusa. Occ. Pap. Nat. Hist. Mus. Stanford Univ., 5:1-90, April 1.

GADOW, H.

1905. Distribution of Mexican amphibians and reptiles. Proc. Zool. Soc. London, 1905:191-244, October 7.

1930. Jorullo. Cambridge Univ. Press, xviii + 100 pp.

GLOYD, H. K.

1948. Description of a neglected subspecies of rattlesnake from Mexico. Nat. Hist. Misc., 17:1-4, April 23.

GOIN, C. J.

1950. Color pattern inheritance in some frogs of the genus Eleutherodactylus. Bull. Chicago Acad. Sci., 9:1-15, March 3.

1954. Remarks on evolution of color pattern in the gossei group of the frog genus Eleutherodactylus. Ann. Carnegie Mus., 33:185-195, June 4.

GOLDMAN, E. A.

1951. Biological investigations in Mexico. Smithsonian Misc. Coll., 115, xii + 476 pp., July 31.

GUZMÁN-RIVAS, P.

1957. Rainfall analysis for some stations in southwest Mexico. In Brand, Coastal study of southwest Mexico, Part I. Univ. Texas, Austin, xii + 140 pp., June 15.

HOROWITZ, S. B.

1955. An arrangement of the subspecies of the horned toad, Phrynosoma orbiculare (Iguanidae). Amer. Midl. Nat., 54:209-218, July.

KLAUBER, L. M.

1945. The geckos of the genus Coleonyx with descriptions of new subspecies. Trans. San Diego Soc. Nat. Hist., 10:133-216, March 9.

1952. Taxonomic studies of the rattlesnakes of Mainland Mexico. Bull. Zool. Soc. San Diego, 26:1-143, August 8.

LANGEBARTEL, D. A.

1959. A new lizard (Sceloporus) from the Sierra Madre Occidental of Mexico. Herpetologica, 15:25-27, February 25.

MALDONADO-KOERDELL, M.

1948. Los colecciones de anfibios del Museo Alfredo Dugès en la Universidad de Guanajuato, I—Urodelos. Mem. Revist. Acad. Nac. Cien., 56:185-226.

MARTIN, P. S.

1958. A biogeography of reptiles and amphibians in the Gómez Farías Region, Tamaulipas, Mexico. Misc. Publ. Mus. Zool. Univ. Michigan, 101:1-102, April 15.

MASLIN, T. P.

1957. Hyla microeximia sp. n., Hylidae, Amphibia, from Jalisco, Mexico. Herpetologica, 13:81-86, July 10.

MILSTEAD, W. W.

1953. Geographic variation in the garter snake, Thamnophis cyrtopsis. Texas Jour. Sci., 5:348-379, September.

MOSIMANN, J. E.

1956. Variation and relative growth in the plastral scutes of the turtle Kinosternon integrum LeConte. Misc. Publ. Mus. Zool. Univ. Michigan, 97:1-43, November 7.

MOSIMANN, J. E., and RABB, G. B.

1953. A new subspecies of the turtle Geoemyda rubida (Cope) from western Mexico. Occ. Pap. Mus. Zool. Univ. Michigan, 548:1-7, November 9.

OLIVER, J. A.

1937. Notes on a collection of amphibians and reptiles from the state of Colima, Mexico. Occ. Pap. Mus. Zool. Univ. Michigan, 360:1-28, November 20.

ORTON, G.

1943. The tadpole of Rhinophrynus dorsalis. Occ. Pap. Mus. Zool. Univ. Michigan, 472:1-8, May 18.

PETERS, J. A.

1954. The amphibians and reptiles of the coast and coastal sierra of Michoacán, Mexico. Occ. Pap. Mus. Zool. Univ. Michigan, 554:1-37, June 23.

1955. Notes on the frog genus Diaglena Cope. Nat. Hist. Misc., 143:1-8, March 28.

1957. The eggs (turtle) and I. Biologist, 39:21-4.

RABB, G. B.

1958. On certain Mexican salamanders of the plethodontid genus Chiropterotriton. Occ. Pap. Mus. Zool. Univ. Michigan, 587:1-37, June 6.

RABB, G. B., and MOSIMANN, J. E.

1955. The tadpole of Hyla robertsorum with comments on the affinities of the species. Occ. Pap. Mus. Zool. Univ. Michigan, 563:1-9, March 29.

REEVE, W. L.

1952. Taxonomy and distribution of the horned lizards genus Phrynosoma. Univ. Kansas Sci. Bull., 34 (pt. 2):817-960, February 15.

SCHMIDT, K. P., and SHANNON, F. A.

1947. Notes on amphibians and reptiles of Michoacán, Mexico. Fieldiana, zool., 31:63-85, February 20.

SHANNON, F. A., and HUMPHREY, F. L.

1957. A new species of Phrynohyas from Nayarit. Herpetologica, 13:15-18, March 30.

1958. A discussion of the polytypic species, Hypopachus oxyrrhinus, with a description of a new subspecies. Herpetologica, 14:85-95, July 23.

SMITH, H. M.

1938. The lizards of the torquatus group of the genus Sceloporus Wiegmann, 1828. Univ. Kansas Sci. Bull., 24:539-693, February 16.

1939. The Mexican and Central American lizards of the genus Sceloporus. Zool. ser., Field Mus. Nat. Hist., 26:1-397, July 27.

1941a. A review of the subspecies of the indigo snake (Drymarchon corais). Jour. Washington Acad. Sci., 31:466-481, November 11.

1941b. Notes on the snake genus Trimorphodon. Proc. U. S. Natl. Mus., 91:149-168, November 10.

1942a. Mexican Herpetological Miscellany. Proc. U. S. Natl. Mus., 92:349-395, November 5.

1942b. The synonymy of the garter snakes (Thamnophis), with notes on Mexican and Central American species. Zoologica, 27:97-123, October 23.

1942c. Remarks on the Mexican king snakes of the triangulum group. Proc. Rochester Acad. Sci., 8:196-207, September 10.

1942d. Descriptions of new species and subspecies of Mexican snakes of the genus Rhadinaea. Proc. Biol. Soc. Washington, 55:185-192, December 31.

1943. Summary of the collections of snakes and crocodilians made in Mexico under the Walter Rathbone Bacon Traveling Scholarship. Proc. U. S. Natl. Mus., 93:393-504, October 29.

1951. The identity of the ophidian name Coluber eques Reuss. Copeia, 1951 (2):138-140, June 8.

SMITH, H. M., and LAUFE, L. E.

1945. Notes on a herpetological collection from Oaxaca. Herpetologica, 3:1-13, November 21.

SMITH, H. M., and TAYLOR, E. H.

1945. An annotated checklist and key to the snakes of Mexico. Bull. U. S. Natl. Mus., 187:iv + 239 pp., October 5.

1948. An annotated checklist and key to the Amphibia of Mexico. Bull. U. S. Natl. Mus., 194: iv + 118 pp., June 17.

1950a. Type localities of Mexican reptiles and amphibians. Univ. Kansas Sci. Bull., 33, pt. 2:313-380, March 20.

1950b. An annotated checklist and key to the reptiles of Mexico exclusive of the snakes. Bull. U. S. Natl. Mus., 199:v + 253, October 26.

SMITH, P. W., and DARLING, D. M.

1952. Results of a herpetological collection from eastern central Mexico. Herpetologica, 8:81-86, November 1.

STICKEL, W. H.

1943. The Mexican snakes of the genera Sonora and Chionactis, with notes on the status of other colubrid genera. Proc. Biol. Soc. Washington, 56:109-128, October 19.

STORM, M.

1939. Hoofways into hot country. Bland Bros., Mexico City, 521 pp.

STUART, L. C.

1943. Comments on the herpetofauna of the Sierra de los Cuchumatanes of Guatemala. Occ. Pap. Mus. Zool. Univ. Michigan, 471:1-28, May 17.

Tamayo, J. L.

1949. Geografía General de Mexico. I—Geografía Fisica. Mexico City, viii + 628 pp.

Tanner, W. W.

1944. A taxonomic study of the genus Hypsiglena. Great Basin Nat., 5:25-92, December 29.

Taylor, E. H.

1936a. New species of amphibia from Mexico. Trans. Kansas Acad. Sci., 39:349-363.

1936b. The rediscovery of the lizard Eumeces altamirani (Dugès) with notes on two other Mexican species of the genus. Proc. Biol. Soc. Washington, 49:55-58, May 1.

1936c. A taxonomic study of the cosmopolitan scincoid lizards of the genus Eumeces. Univ. Kansas Sci. Bull., 23:1-643, August 15.

1939a. Concerning Mexican salamanders. Univ. Kansas Sci. Bull., 25:259-313, July 10.

1939b. Frogs of the Hyla eximia group in Mexico, with descriptions of two new species. Univ. Kansas Sci. Bull., 25:421-445, July 10.

1940a. A new bromeliad frog from northwestern Michoacan. Copeia, 1940 (1):18-20, March 30.

1940b. Mexican snakes of the genus Typhlops. Univ. Kansas Sci. Bull., 26:441-444, November 27.

1940c. Some Mexican serpents. Univ. Kansas Sci. Bull., 26:445-487, November 27.

1940d. Herpetological Miscellany. Univ. Kansas Sci. Bull., 26:489-571, November 27.

1941a. Some Mexican frogs. Proc. Biol. Soc. Washington, 54:87-94, July 31.

1941b. Herpetological Miscellany, No. II. Univ. Kansas Sci. Bull., 27:105-138, December 30.

1942a. New tailless Amphibia from Mexico. Univ. Kansas Sci. Bull., 28:67-89, May 15.

1942b. Tadpoles of Mexican Anura. Univ. Kansas Sci. Bull., 28:37-55, May 15.

1942c. The frog genus Diaglena with a description of a new species. Univ. Kansas Sci. Bull., 28:57-65, May 15.

1943a. Herpetological novelties from Mexico. Univ. Kansas Sci. Bull., 29:343-361, October 15.

1943b. A new Hylella from Mexico. Proc. Biol. Soc. Washington, 56:49-52, June 16.

1952. A new hylid frog of the genus Agalychnis from southwestern Mexico. Copeia, 1952 (1):31-32, June 2.

1956. A review of the lizards of Costa Rica. Univ. Kansas Sci. Bull., 38 (pt. 1):3-322, December 20.

TAYLOR, E. H., and SMITH, H. M.

1942a. Concerning the snake genus Pseudoficimia Bocourt. Univ. Kansas Sci. Bull., 28:241-249, November 12.

1942b. The snake genera Conopsis and Toluca. Univ. Kansas Sci. Bull., 28:325-363, November 12.

1945. Summary of the collections of amphibians made in Mexico under the Walter Rathbone Bacon Traveling Scholarship. Proc. U. S. Natl. Mus., 95:521-613, June 30.

TIHEN, J. A.

1949. A review of the lizard genus Barisia. Univ. Kansas Sci. Bull., 33:217-256, April 20.

VIVÓ, J.

1953. Geografía de México. Mexico City, ed. 3, 338 pp.

WEBB, R. G.

1958. The status of the Mexican lizards of the genus Mabuya. Univ. Kansas Sci. Bull., 38:1303-1313, March 2.

WELLMAN, J.

1959. Notes on the variation in and distribution of the Mexican colubrid snake Coniophanes lateritius. Herpetologica, 15:127-128, September 10.

WOODBURY, A. M., and WOODBURY, D. M.

1944. Notes on Mexican snakes from Oaxaca. Jour. Washington Acad. Sci., 34:360-373, November 15.

ZWEIFEL, R. G.

1957. A new frog of the genus Rana from Michoacán, Mexico. Copeia, 1957 (2):78-83, July 15.

1959a. Variation in and distribution of lizards of western Mexico related to Cnemidophorus sacki. Bull. Amer. Mus. Nat. Hist., 117:57-116, April 27.

1959b. Additions to the herpetofauna of Nayarit, Mexico. Amer. Mus. Novitates, 1953:1-13, June 26.

1959c. Snakes of the genus Imantodes in western Mexico. Amer. Mus. Novitates, 1961:1-18, September 16.

1960. Results of the Puritan-American Museum of Natural History Expedition to western Mexico. 9. Herpetology of the Tres Marías Islands. Bull. Amer. Mus. Nat. Hist., 119:77-128, February 29.

1961. Relationship of two whiptail lizards (genus Cnemidophorus) in western Mexico. Copeia, no. 1:98-103, March 17.

Lector House believes that a society develops through a two-fold approach of continuous learning and adaptation, which is derived from the study of classic literary works spread across the historic timeline of literature records. Therefore, we aim at reviving, repairing and redeveloping all those inaccessible or damaged but historically as well as culturally important literature across subjects so that the future generations may have an opportunity to study and learn from past works to embark upon a journey of creating a better future.

This book is a result of an effort made by Lector House towards making a contribution to the preservation and repair of original ancient works which might hold historical significance to the approach of continuous learning across subjects.

HAPPY READING & LEARNING!

LECTOR HOUSE LLP
E-MAIL: lectorpublishing@gmail.com